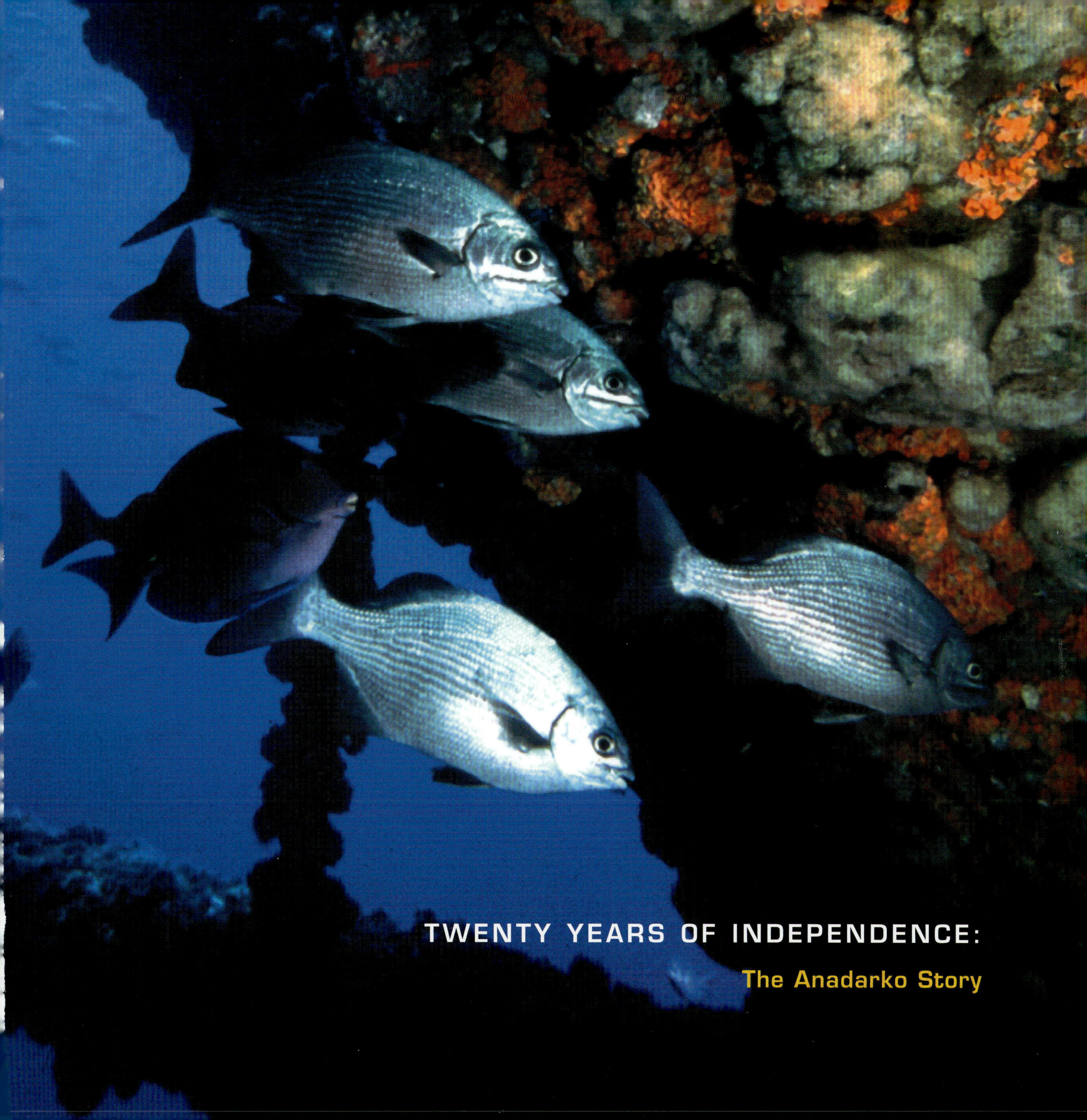

TWENTY YEARS OF INDEPENDENCE:

The Anadarko Story

TWENTY YEARS OF INDEPENDENCE:

The Anadarko Story

BY RUSS BANHAM

GREENWICH PUBLISHING GROUP, INC.

Produced and published by Greenwich Publishing Group, Inc.
Old Saybrook, Connecticut
www.greenwichpublishing.com

Design by Clare Cunningham Graphic Design

Library of Congress Catalog Card Number: 2006930858
ISBN: 0-944641-68-7

First Printing: November 2006

10 9 8 7 6 5 4 3 2 1

Photo Credits:
Page 15: © Louie Psihoyos/CORBIS
Pages 32 (upper left), 33 (both), 34 (upper left), 35 (lower right), 37 (both), 38 (lower left) and 39 (lower left) appear courtesy of Woodson Research Center, Fondren Library, Rice University

Anadarko would like to express its sincere thanks to the following people who have helped document Anadarko's history and milestone achievements through photography. They include: John Becq, Jacek Bogucki, Melanie Campbell-Tello, Ken Childress, Jeff Cline, Rich Frischman, Jay Rusovich, John Smallwood, Greg Smith, David Tejada and Teresa Wong.

All other photographs and historical items appear courtesy of Anadarko Petroleum Corporation and its subsidiaries.

TABLE OF CONTENTS

CHAPTER ONE: RESOURCES, NATURAL AND HUMAN

Longtime Explorer

Lee Petersen, left, studied to be a paleontologist and leveraged his knowledge of rocks toward the hunt for fossil fuels. Over a career span of almost 30 years with Anadarko, beginning as a young man and today as Anadarko's worldwide exploration advisor, Petersen has been passionate about his work.

Lee Petersen found his calling as a geologist as a five-year-old boy. His paternal grandmother shuttled him a couple times a month to Chicago's Museum of Natural History, where his youthful imagination was stirred "by dinosaurs and rocks," he says. His seventh-grade science teacher Robert Ferguson cultivated his budding interests, as did his high-school teachers. Lee learned about and collected different kinds of rocks and, on occasion, headed into the hills on field trips with his father, a commercial artist, to dig up fossils.

Chester Johnson, a one-man geology department at Gustavus Adolphus, a small liberal arts college in Minnesota, proffered further guidance to Lee's chosen path. Petersen graduated with degrees in biology and geology, and planned to become a paleontologist, studying forms of life that existed in different geologic periods. On a fellowship at Arizona State University, Lee earned master's and PhD degrees in geology with a specialization in paleontology. To support himself and his wife Virginia and their two children, he worked as a summertime guide on the Colorado River.

Petersen hoped to teach at the graduate level, but there were few teaching jobs available, particularly in paleontology. He instead turned his sights to the oil patch. He got a job at Cities Service Oil and Gas in Jackson, Mississippi, generating promising oil and gas prospects in East Texas, Northern Louisiana and Mississippi. Three years later, he went to work for tiny Coquina Oil and Gas in the same capacity. In 1978, while reading the newspaper, Petersen spied a hiring notice placed by Anadarko Production Company in Denver. He applied and won a job in exploration, launching his long career with Anadarko in the Hugoton gas field.

Petersen is one of the many thousands of people who over the years have worked at Anadarko Petroleum Corporation, long a worldwide leader in oil and gas exploration and production, whose success over five decades has been built less on natural resources than on those of the human kind.

Anadarko's current workforce of 5,300 is, in fact, a portrait of humanity, an album of people from all walks drawn to the industry along different paths. Some studied rocks or picked apart watches as children. Some were told they had a distinctive aptitude for math or science. Others were reared in regions rich in hydrocarbon deposits, the seed core of much of the world's energy. For them, the oil and gas industry was the town's major employer, the home away from home of fathers, uncles and aunts, an exciting place where words like "gusher," "roughneck" and "roustabout" stirred youthful imaginations. When they grew up, they, too, wanted the life.

For Petersen, what a career it has been — full of an incredible variety of projects, from the subsalt Gulf of Mexico to the Alpine field in Alaska. Lee's proudest moment was Anadarko's first great oil find in Algeria, a "make-or-break" event for the company. "I ran into one of our senior financial people in the parking lot one day and he said, 'You know, Lee, Anadarko really needs this, and if it doesn't hit it's all over,'" says Petersen, Algeria exploration manager at the time. "We had just started drilling. I think I gulped."

Although Petersen retired in 2002, Anadarko kept calling him back as a consultant on different assignments. Today, he is back fulltime as Anadarko's worldwide exploration adviser. "Either they can't let me go or I haven't ticked them off enough yet," he laughs. "I am passionate about this company."

Built at the Drill Bit

Anadarko's primary objective is to develop, acquire and explore for oil and gas resources vital to the world's health and welfare. The company has discovered some of the world's largest oil and gas deposits, yet it began humbly in 1959 as the subsidiary of a natural gas utility, Panhandle Eastern Pipe Line Company. Panhandle formed Anadarko to get out from under government regulations that prohibited pipeline companies from selling oil and gas for anything above cost of service. As a production company, Anadarko

Persian Gulf Presence

Huda East, above, is executive assistant to the president of Anadarko Qatar Energy Company. Anadarko's 2001 acquisition of GulfStream Resources gave it ownership of three blocks offshore Qatar, including a 19,000-acre block at the Al Rayyan field. Qatar's scenic beauty encompasses palm tree-lined beaches, left, and colorful markets called souks, above opposite.

executive assistant to the president, rattles off the nationalities of her 30 fellow workers. "We have Filipino, Nepalese, Qatari, Tunisian, British, Welsh, Irish, Pakistani, Jordanian, Palestinian, American, Arabs of different nationalities, and I'm Sudanese," she says. "The mix of different cultures and languages can be quite a lot of fun."

Anadarko is among the largest acreage holders in Qatar, with more than 1.6 million gross acres under lease for exploration. Doha has developed over the past 30 years from a quiet pearl-trading and fishing town into a thriving commercial, Westernized city, yet its people have retained their traditional Arab values of hospitality and generosity. "The open and friendly atmosphere of the *souks* (local markets) remains," says East. "The people are so welcoming. It has been an ideal place to bring up my three children, who are grown now."

East's native tongue is Arabic. She also speaks perfect English, which she learned in a Sudanese convent. Her bilingual skill earned her jobs at several oil and gas companies in Libya and Qatar before she joined Anadarko in 2002. Her responsibilities include community affairs and training programs for new employees. She also heads up Anadarko's "Qatarization" program — a requirement by the Oil Ministry to staff at least half of the workforce of companies doing business in Qatar with Qatari nationals. "We're at 35 percent now, well ahead of most other companies," she says.

would be free to sell its resources for market rates and make a profit. Hence, exploration and production have been Anadarko's *raison d'etre* from the very beginning.

Major growth ensued in the years following Anadarko's spinoff from Panhandle as a public company in 1986. Anadarko pursued an aggressive exploration program, drawing upon its staff's technical competence to build the company through the drill bit. It further parlayed expertise in unconventional technology and processes to coax oil and gas from long-dormant fields essentially forsaken by the rest of the industry. Leveraging its diverse skills in the rich oil and gas basins of the American heartland, the Gulf of Mexico and the Algerian desert, Anadarko today is one of the industry's largest international explorers, with significant holdings in North America and exploration and production in North Africa and the Middle East.

At the headquarters of Anadarko Qatar Energy Company in Doha, Qatar, Huda East,

"Sometimes You Work with Your Heart"

When Ade Adeleye's parents — an Army officer and a teacher — learned that their son was skilled at math, they steered him toward science and engineering. He graduated from college in his homeland of Nigeria, having been born and reared in the southwestern part of the African nation. He then moved on to earn a master's degree in petroleum engineering at Heriot-Watt University in Edinburgh, Scotland. He was recruited and hired by BP (British Petroleum) and later traveled the globe on behalf of Shell Oil, which hired him in its North Sea drilling group. "Everything was planned, not that I have any regrets," Adeleye says.

Today Adeleye is a senior staff drilling engineer at Anadarko's headquarters in Houston. He and two colleagues developed a 10-step process that improved the drilling efficiency of Anadarko's Wild River *play* — an oil-industry term denoting a promising geologic trend. "We were able to reduce drilling days from between 27 to 40 days in 2002 to between 14 and 22 days, which cut overall drilling costs in half," Adeleye notes. The company has since applied the process to several other areas, and Adeleye has turned his attention to improving deepwater drilling efficiency technology. "I love finding new ways to attack difficult challenges."

Aziz Abad also has seen much of the world, initially as a marketer of crude oil for Sonatrach, the state-owned Algerian oil and gas concern. Today, Abad works in the London office of Anadarko as its international crude oil marketing manager, selling Saharan Blend — the light, sweet oil produced jointly by Anadarko and Sonatrach in the Sahara Desert. From 1989 to the present, Anadarko explorationists have discovered nearly three billion gross barrels of

Efficient Drilling Technology

Ade Adeleye, an Anadarko senior staff drilling engineer, worked with colleagues to improve productivity and substantially lower drilling costs. He helped pare drilling costs in half at the Wild River play in western Canada, above, before turning his attention to technology to improve deepwater drilling efficiency.

The Road to Convergence

oil and condensate in Algeria. These days, roughly 500,000 barrels of oil are produced daily. "Because it is high-gravity and low-sulphur, Saharan Blend yields high-quality refined products such as diesel fuel that is in high demand in Europe," says Abad. Sales recently have spread to the United States and Far East.

Abad speaks English, Arabic, Berber and French "with an Algerian dialect," he says. "I come from a family of Berbers, the people who lived in Algeria for centuries prior to the invasion by France. When I was 19 years old, I was sent by Sonatrach to take a course in chemical engineering in the U.K. I didn't know a word of English then, but I'm a fast learner." He considers communication key to marketing crude oil. "I find it helpful to try to speak the language of the people I'm dealing with and to try to absorb their culture," he explains. "Sometimes you have to work with your heart as well as with your brain."

No Ordinary Job

Many members of Anadarko's management team got their start in the industry as roustabouts, people with demanding physical jobs like hauling pipe or supplying equipment to a rig. One of them is Kendall Madden, Anadarko's foreman at the Bossier field in East Texas. Madden, an Oklahoma native, is in charge of the field's physical operations. "We take it from

Light, Sweet Saharan Blend

Anadarko hit the mother lode in the Saharan sands of Algeria, discovering nearly three billion gross barrels of oil and condensate. Production is processed at central processing facilities, such as the one opposite. As Anadarko's international crude oil marketing manager, Aziz Abad sells the prized Saharan Blend, a versatile grade of oil Anadarko produces jointly with Sonatrach, the state-owned oil company of Algeria.

Squeezing Gas from Tight Sands

Gas production in the East Texas Bossier play was uneconomical until Anadarko found a way to break through the field's tight rock with water-based hydraulic fracturing. Kendall Madden, opposite at left, is foreman at the field, where almost 30 rigs have run at a time. Ken Filardo Jr., at right opposite, is foreman at Louisiana's Vernon field, a play made successful by the technology expertise gleaned at Bossier.

stake to first sales," he explains. "In 1997, we had seven wells and two rigs running. At our peak, we had 28 rigs going and were producing 375 million cubic feet of gas a day. To date, we've drilled more than 700 wells in the Bossier and have more than 500 miles of pipeline in our gathering system."

As a young man, Madden joined the oil patch "at the other end of a shovel," he says. "It was about the only job you could get then with no experience. Later on, I laid pipeline during the big oil boom in the late 1970s, but I don't do physical labor anymore. I've got a first-rate group of people I manage now, men and women who like their jobs because of the challenge. About the most routine thing any of us do is get out of bed each morning. Every day, we're exposed to something different here."

The Bossier field is a bounteous natural gas basin bedeviled by what geologists call "tight gas sands," which make gas difficult to extract from the stingy basin. The company has liberated the gas by cracking the rock with water injected at high pressure, a process known as hydraulic fracturing — one of several enhanced recovery technologies at which Anadarko is particularly adept.

Ken Filardo Jr. is Madden's counterpart at the Vernon field in Louisiana, which also contains tough-to-harvest deposits of gas. At Vernon, the company leveraged the knowledge and expertise it picked up at Bossier. Filardo grew up in Houma, Louisiana, and found his first job in the industry working for a catering service that delivered food and beverages to employees working in the Gulf of Mexico. He came to like the offshore business and secured a position as a roustabout while studying to become a production foreman. He took college classes in petroleum engineering, but when the oil patch thrived

with job opportunities in the early 1980s he left school to work with Union Pacific Resources. "Every day is a learning experience," he says. "There is always this great sense of pride and excitement in our work."

Julie Struble can relate to such feelings. As a child in the small town of Crowley, Louisiana, Struble had "a real affinity for math," she says, guiding her to pursue a degree in chemical engineering at Louisiana Tech. "When I went through a course on petroleum engineering, I was hooked," she says. "I thought, 'This is it. This is what I want to be.'"

Although there were few women in the oil patch upon her graduation in 1982, she eschewed preferential treatment. "I was in the upper percentile of my class and I wanted a company that was interested in me as an engineer first and foremost," she explains. "What drew me to Anadarko were these down-to-earth, incredibly intelligent people who treated me as a professional." Struble steadily ascended the corporate ladder and today is vice president, operations – southwestern region. "I've had so many great opportunities," she says. "I'm pleased with where I am and what lies in front of me."

In Good Company

The West Texas Permian Basin often is referred to as Anadarko's "onshore subsalt play" because of its 5,000-foot section of evaporite — seismic-defying sediment that remains after the salty water in which it had dissolved evaporates. Julie Struble, opposite, manages operations at the basin's Haley field, right. She was one of the industry's few female petroleum engineers when she started working for Anadarko in the early 1980s.

In 2006, two strategic acquisitions by Anadarko — Kerr-McGee Corporation and Western Gas Resources — delivered hundreds of new people into the company's embrace, such as Kerr-McGee Senior Consultant Beth Kendall. Born and reared in Boulder, Colorado, Kendall was "kind of a science whiz kid" growing up, though she was unsure where her talents would lead. Then, in high school, "we had career day, and after the accountants and lawyers did their speeches, a couple of geophysicists came in and put together this puzzle of what the rocks looked like under the Earth," she says. "I thought it was the coolest job … and knew this was what I wanted to do with my life."

She graduated from the Colorado School of Mines and hired on as a geophysicist at Atlantic Richfield, leaving after 17 years to work for Oryx Energy Company, which was bought by Kerr-McGee in 1998. Her career highlight was sharing the discovery of the Constitution reservoir in the Central Gulf of Mexico, "talking the company into drilling it, and then not giving up when the first well was disappointing," says Kendall. The persistence of Kendall and her team paid off with a 100 million-barrel field.

Constitution Spar

Anadarko geophysicist Beth Kendall, formerly of Kerr-McGee, and geologist Paul Jaeger are credited with the discovery of the deepwater Constitution field in 2001. When it came time to christen the Constitution spar in Pori, Finland, Kendall was given the honor of hoisting and shattering a bottle of champagne against the hull of the spar, shown above at work in the Gulf of Mexico.

EasyCopy - center.png
File Edit View Capture Image Help
SPAR FEATURES
Constitution
Capacity
Spar Diameter
Spar Length
Topsides Payload
Hull Weight
Location
Water Depth
Dry Tree Slots
Subsea Tieback Pairs
Gas Export
Oil Export
Workover Rig
70,000 BOPD
200,000 MMcf/d
98 Feet
554 Feet
10,770 Tons
14,000 Tons
Green Canyon 680 GOM
4,970 Feet
14" SCR
14" SCR
API 1500 HP
CONSTITUTION
Benefits of Truss
TOPSIDES

Mountain Methane

Dave Keanini is Anadarko's general manager of engineering for the Rocky Mountain region. His previous job as senior process engineer at Western Gas Resources was critical in developing the company's coalbed methane reservoirs in Wyoming's Powder River Basin, where Western Gas was one of the largest acreage holders and producers.

Dave Keanini also graduated from the Colorado School of Mines, in his case with a degree in chemical engineering. As a boy in Golden, Colorado, Keanini liked hiking to the top of Lookout Mountain, where Wild West showman Buffalo Bill Cody is buried. Afterwards, he'd stop by the School of Mines to play pinball. "I'd like to say I was drawn to the school by its chemical engineering program, but it was really the game room," Keanini laughs. Leveraging a full-ride academic scholarship from Coors Brewing Company, where his dad worked in the ceramics plant, Keanini graduated and took a job as a process engineer at Colorado Interstate Gas. He found his way to Western Gas Resources in 1989 and was the project manager on the company's breakthrough Reno Junction butane isomerization project in Wyoming. "We essentially took field-grade butane and converted it into higher-market-value iso-butane," says Keanini, general manager of midstream engineering for the Rockies at Anadarko. "It was the biggest project we had done at the time."

Anadarko employees like Lee Petersen, Huda East, Ade Adeleye, Aziz Abad, Kendall Madden, Ken Filardo, Julie Struble, Beth Kendall, Dave Keanini and thousands of others are bound by a set of values, such as acting with integrity, valuing people as the competitive edge, learning and continuously improving, focusing on sustainable commercial success and building trust with stakeholders. Over the past 50 years, many of them have explored for oil and gas and produced it in some of the harshest conditions on the planet, from the sub-zero temperatures of Alaska's North Slope to the flat plains of the dust bowl, from the deepest waters in the Gulf of Mexico to the unforgiving sunshine of the Sahara. They go to work every day knowing that theirs is no ordinary job and that the energy of the world literally depends on them.

CHAPTER TWO: "WE WERE ON OUR OWN"

In the Beginning

Missouri-Kansas Company founder Frank Parish, above, dropped out of school at age 14 and worked as a farm laborer, teamster, machinist and store clerk. Often penniless, he hitched rides on trains from one job to the next. At 21, he formed a company to manufacture mine-sweeping equipment and made a small fortune in sales to the U.S. military during World War I. Parish's later work as a pipe salesman led him in the 1920s to the oil patch, right, still in its infancy. Mo-Kan's internal magazine, *Mo-Kan Bulletin*, reported on the work of early seismic crews, opposite.

It was a hazy, spring morning in 1959 in the small town of Liberal, Kansas, when field office personnel of the sprawling Panhandle Eastern Pipe Line Company were notified that they now had two jobs. They'd be working for the 31-year-old natural gas pipeline company as well as for a new exploration and production subsidiary. "They came around and said that we're going to get a paycheck and it's going to have 'Anadarko Production Company' on it," the late Tom Holland, a Panhandle land advisor at the time, once recalled. "'Who's that?' I said."

The unknown entity with the unusual name would grow rapidly over the ensuing decades to become one of the world's largest independent oil and gas exploration and production companies. But at 9 a.m. on June 8, 1959, the time and day when Anadarko was officially incorporated in the state of Delaware, the subsidiary was little more than a strategy to boost sagging profits. "Panhandle told us we were going to handle the world from Liberal, Kansas," Holland said in a 1989 interview. "And we thought, 'By golly we can!'"

Panhandle Eastern had been formed in 1928 as Missouri-Kansas Company, known colloquially as Mo-Kan. The company owned modest reserves in the vast Hugoton and Panhandle hydrocarbon fields buried beneath the high plains of southwestern Kansas and the Oklahoma and Texas panhandles. Contracts to buy natural gas produced from these fields were arranged with local utilities, but as Mo-Kan searched for additional markets it aroused the ire of the Power Trust, the collection of large utility monopolies that controlled the flow of gas in a territory stretching from the Midwest to the Northeast.

In the Roaring Twenties, railroad barons, shipping tycoons, oil magnates and utility titans

controlled a vast expanse of commerce in the United States, earning billions of dollars to become America's ruling commercial class. Largely unencumbered by regulatory constraints, powerful men created indomitable monopolies that manipulated markets and crushed the hopes of competing entrepreneurs. Their dominance dared few to enter their domains. Frank Parish, the founder of Panhandle Eastern, dared back.

When the Power Trust closed off markets in the north, Parish formed more than a dozen subsidiaries to develop reserves and produce and transmit natural gas in Indiana, southwestern Kansas, Kentucky and the Texas and Oklahoma panhandles. By 1930, Mo-Kan controlled leases on 500,000 gas-producing acres across seven states, boasting 1.5 trillion cubic feet of natural gas reserves in the Hugoton field alone. But Parish needed customers for his gas and a way to deliver the fuel to them.

He conjured a radical, "build it and they will come" idea: a 1,000-mile interstate pipeline originating in the Hugoton field and terminating in the Twin Cities of Minnesota. Knowing that the Power Trust would do all it could to block the route, an associate, William G. Maguire, suggested they end the pipeline in Indianapolis instead. Not only was that city in the midst of rapid industrialization, it was close to other burgeoning markets thirsty for natural gas as an alternative to the day's conventional form of energy, gas derived from coal. The route change also appealed to North American Light and Power, an Illinois-based utility that agreed to purchase Mo-Kan gas. With the contract firmly in hand, Parish raised capital in the Chicago and New York financial markets to build the world's largest natural gas transmission system.

The new eastward route of the pipeline and the source of its supply in the Texas Panhandle compelled Parish to change the name of Mo-Kan to Panhandle Eastern Pipe Line Company in 1930. The pipeline's construction kicked off that March at an estimated $40 million cost. Liberal, located in the southwestern part of Kansas, was selected as the site of the main compressor station. Ultimately the pipeline's terminus was extended to the industrial nexus of Detroit, home to the budding automotive industry, and was connected to a pipeline owned by Columbia Gas and Electric to provide natural gas to the Atlantic seaboard.

The Birth of Anadarko

From the 1930s through the 1950s, Panhandle Eastern endured enormous strains in the face of significant growth. The company funded the building of another major gas transmission system, the 1,300-mile Trunkline pipeline, to ferry gas reserves from the Texas and Louisiana coasts to the Panhandle system in the Midwest. The

Natural Gas Act of 1938

Mo-Kan built its business foundation on natural gas, which *Fortune* magazine described in 1940 as "clean as air and as convenient as tap water." In the Hugoton field alone, the company boasted approximately 1.5 trillion cubic feet of natural gas reserves. The Natural Gas Act of 1938, signed by President Franklin D. Roosevelt, right, crimped plans to further explore for new gas reserves.

development of Panhandle and other gas transmission companies was constrained, however, by stringent federal regulations. The Natural Gas Act of 1938 gave the Federal Power Commission (FPC), the forerunner of today's Federal Energy Regulatory Commission, the authority to regulate the sale and transmission of natural gas. The law replaced a decades-old system in which states had regulated intrastate gas sales and interstate sales had gone unregulated.

The noose tightened further in 1954 when the U.S. Supreme Court ruled that Congress, in passing the Natural Gas Act, also had intended to control the price at which producers sold gas to pipelines. The federal government could now impose ceilings on the price of natural gas. Christopher Castaneda, co-author of the book *Gas Pipelines and the Emergence of America's Regulatory State* and a professor of business history at California State University, explains that the Federal Power Commission "was very consumer-oriented, and would keep natural gas prices low through the 1950s and into the 1960s, ultimately to the detriment of both the market and consumers."

The FPC effectively prohibited Panhandle and other companies that owned pipelines from selling oil and natural gas for anything above cost of service. With a cap on prices, there was little financial incentive for a pipeline company to explore for new gas reserves. "We didn't want to give up exploration and production activities, but it became obvious a pipeline company that owned production assets wasn't going to be afforded the same treatment as other producers," says Richard O'Shields, CEO and chairman of Panhandle Eastern from 1970 to 1983.

In a bind to increase profits, Panhandle diversified. It produced helium and petrochemicals and refined crude oil. In 1959, Maguire, who had succeeded Parish as CEO in 1943, fashioned a better way to raise profits: He created a subsidiary designed strictly for exploration and production and then transferred Panhandle's undeveloped properties in the Anadarko Basin to that concern.

Anadarko Production Company was born.

As an independent producer with no pipeline interests, Anadarko could develop its properties free from low FPC-regulated cost-of-service returns. Gas produced could be sold at the fair field price — about 10 cents per thousand cubic feet at the time, much higher than gas sold by pipeline companies. "Those undeveloped leases in the Anadarko Basin were a great asset, but the only way stockholders were going to realize the value of those assets was to get them out from under pipeline-type regulation," R. C. Dixon, Panhandle's manager of gas purchasing in 1959, said in a 1989 interview.

The name Anadarko derived from the Anadarko Basin in the Texas and Oklahoma

panhandles and southwest Kansas, an immense oil and gas field named for the Ahnadakka Indians, a tribe that originally lived in and around Nacogdoches in East Texas. Following Anadarko's incorporation, Panhandle purchased Anadarko's common stock for roughly $4 million — $501,000 in cash and the balance in the value of 544,985 acres of exploration properties in the Anadarko Basin. Liberal, Kansas, site of the compressor station for the great Panhandle pipeline, became Anadarko's headquarters.

The new company was staffed largely by people from Panhandle's exploration and production departments. "We wore two hats," recalls

Shipping by LNG Tanker

Trunkline Gas Company, a Panhandle Eastern subsidiary, built the first extensive natural gas supply system in the Gulf of Mexico in the 1950s and 1960s. The company was the first to deploy a 30-inch diameter pipe in the Gulf. Liquefied natural gas ships, built by companies like General Dynamics, provided an alternative to pipelines to move natural gas great distances. An LNG tanker, above, unloads in shallow waters on the shore of Lake Charles, Louisiana.

Anadarko's first logo capitalized on its unique heritage — the Ahnadakka Indians, represented by the tepee, and the oil industry, represented by the derrick. The tribe was renowned for agriculture, planting more than 150 acres of "some of the finest corn ever seen in Texas," according to University of Texas historians.

What's in a Name?

Robert B. Harkins, Anadarko's first official president, is credited with naming the company. "Anadarko" was a natural name, as most of the company's properties were located in the Anadarko Basin in Oklahoma, which also boasts a town called Anadarko — both named after the Ahnadakka Indians, a division of the Caddoes, who lived along the Red River in Arkansas, Texas, Louisiana and Oklahoma. Harkins also had an affinity for the Native American heritage, so the company captured these tribal origins in its first logo — the word "Anadarko" set within a tepee.

The tribe, whose name is alternatively written as "Anadaca," "Anduico," "Nandacao" and "Nadaco," apparently had a fondness for eating honey — "nadarko" is a centuries-old Indian word signifying "those who ate the honey of the bumble-bee." These Native Americans lived originally in the Nacogdoches and Sabine River areas of East Texas. When the Texas legislature voted in 1854 to establish reservations for the state's remaining Indians, the Ahnadakka were moved to a reservation on the Brazos River. Five years later, they were relocated again, about 400 miles north to a reservation in the area of present-day Anadarko, Oklahoma.

The Ahnadakka were made famous by their Chief Iesh, who was born in 1805. Recorded in history books as Jose Maria, the Christian name given him by Spanish missionaries, the chief was a champion peacemaker between white settlers and Indian tribes in Texas. He helped the Texas Rangers bring order to the frontier and defend it against hostile tribes and white renegades. America's 11th president, James K. Polk, honored the chief's peacemaking endeavors and friendship in the 1840s.

Gary Cobb, a Panhandle reservoir clerk at the time.

Anita Simon, then a stenographer in the land department, remembers the overtime she received dividing two companies' records. "I'd do a contract for Panhandle one minute and then the next thing I'd do was a contract for Anadarko," she says. "I bought a car with that overtime."

Cobb, Simon and about 50 other employees worked in a large, one-story, curved building that Panhandle had built just a few months earlier, having moved its offices from a former dime store named Duckwalls. They were easily recognizable in the small agricultural town of a few thousand, decked out in business suits and ties or modest dresses on even the most sweltering summer days. Many were excited about the company's future. "We were all young and we saw an opportunity to do it the way we'd always dreamed," said the late Tom Holland.

Frederick H. Robinson was Anadarko's first president, but it was a titular position. "We had to have some people's names put down for the principal officers as part of the incorporation," O'Shields explains. "The operations were really handled by Bob Harkins," vice president for gas supply and production at Panhandle. Robert B. Harkins, officially named Anadarko president in 1962, was a serious and energetic man. One employee at the time joked that Harkins' initials stood for "real big hurry."

Holland said, "Bob was a workaholic. He'd call a meeting on the Fourth of July. He'd call a meeting anytime. We'd go to work on Saturdays and in blizzards, but we enjoyed each other and we enjoyed the company."

Harkins' emphasis was on growth, which he built on forays for oil and gas beyond Panhandle's legacy assets. "We were buying up a lot of land and were leasing like mad in places like Arkansas

Father of Anadarko

William G. Maguire was a respected Midwest businessman with long ties to the utility industry. As Mo-Kan's advisor in its early years, his contacts were crucial in negotiating gas sales contracts with major municipalities, utilities and agricultural equipment makers. Later, as Panhandle Eastern CEO, he established Anadarko Production Company as a subsidiary. Maguire turned on the gas at the Trunkline pipeline at the official opening ceremony, above, in 1951.

and Colorado," Simon recalls. "I remember the landmen went into the remote, back hill country of Arkansas carrying a manual typewriter to type the leases on. They found a local notary who knew the landowners, many of them concerned that the landmen were revenuers looking to collect taxes. When they were told they'd be paid money to drill on their land, most of them signed, a few with an 'X.'"

By the end of 1959, Anadarko had participated in drilling 17 wells in the Anadarko Basin, achieving notable success: One of three exploratory wells and all 14 development wells were producers. The first well spud was the Harris 1-5 in Meade County, Kansas, on August 10, 1959. It was connected to Panhandle's transmission line the following May and was plugged in 1971 after producing 591 million cubic feet of natural gas and 2,000 barrels of condensate.

During its first full year of operations, 1959, Anadarko increased its participation in drilling to 48 wells, spending $2.5 million on exploration. The company purchased 27 producing wells in the Texas Panhandle and leaseholds on 182,000 acres in the Anadarko Basin, western Oklahoma and southeast Colorado. "We were always pushing hard," says Ellen Cobb, who started her career at Panhandle in 1955 in Liberal and retired as a senior marketing analyst with Anadarko in 2003. She and Gary Cobb, who retired as a computer systems analyst with Anadarko in 1999, have been married for more than 50 years.

Expansionist CEOs

Anadarko CEOs Richard O'Shields, left, and R. C. Dixon, opposite below, led the company during the 1960s and early 1970s, a time when demand for oil and gas outpaced the supply. To boost production, O'Shields added offshore Gulf of Mexico exploration to development of major onshore basins. Dixon pursued a similar strategy, decentralizing the company into regional offices.

Target: Expansion

Anadarko entered into its first long-term contract in 1960. It was a 20-year agreement with Amarillo's Pioneer Natural Gas Company to provide gas from the Red Cave formation in the Texas Panhandle. Anadarko built its first pipeline, an 84-mile intrastate line in Kansas to supply factories in the Wichita area. Through the early sixties, it continued to develop properties in the Anadarko Basin, increasing gas production from 27 billion cubic feet in 1962 to more than 53 billion cubic feet in 1964. Oil production over the same span doubled from 911,000 barrels to 1.8 million barrels. To grow into a more substantial enterprise, O'Shields says, Anadarko "had to expand into more oil and gas provinces than just the Anadarko Basin."

That objective was accomplished in 1965 when Anadarko purchased the assets of Ambassador Oil Corporation of Fort Worth, Texas, for $12 million. The assets included working interests in proven oil and gas reserves in the Arkoma Basin in east-central Oklahoma and west-central Arkansas, and undeveloped leases on about 600,000 acres in 19 states and Canada. This more than doubled Anadarko's lease position to over one million net acres. Ambassador also had interests in Italy, England and Ireland. "At one time, Ambassador actually had exploration rights to the entire island of Ireland," says O'Shields, who notes that the main purpose of the acquisition was to build up Anadarko, which by then was contributing handsomely to Panhandle's profits.

Concluding the acquisition, Anadarko relocated its headquarters from Liberal to Fort Worth, close to the hub of the U.S. petroleum industry. The company assumed Ambassador's former offices on the first floor of a two-story building on Camp Bowie Street and absorbed most of its personnel. "We had about 100 people in the company back then," says Simon. "You could still remember everyone's names."

Anadarko inherited more than Ambassador's assets and employees in the acquisition. It also picked up some squawky Hollywood celebrities as shareholders. James Stewart, James Dean, Marlon Brando, Ann-Margaret and Alfred Hitchcock had been among the leading investors in wells drilled by Ambassador in the 1950s; Stewart held the biggest working interest in the company's prospects in the Arkoma Basin, followed closely by Brando.

O'Shields says Anadarko tried to buy most of them out at the time of the acquisition, but that a few — apparently infatuated with the boom-or-bust oil business — "wouldn't sell or sign anything at all for years." In the meantime, he says, the stars seemed all too willing to take their share of the profits when a prospect panned out but all too reluctant to fork over their share of the operating costs. "When they had to pay up, some wouldn't," he says. "The only way we'd get money out of them was by withholding the proceeds when production was up until they paid their share." Eventually all of the stars sold their shares to Anadarko.

Anadarko Restructures

Harkins was the mastermind behind the acquisition, but he died before the contracts were signed. Dick O'Shields, tall, intelligent and distinguished-looking, succeeded Harkins as president in 1965. It was evident that demand for oil and gas was outpacing the rate of new discoveries, and signs on the horizon indicated severe energy shortages. O'Shields' strategy was to continue to develop oil and gas reserves in the major basins and to bolster them with offshore exploration and production in the Gulf of Mexico. In 1968, he was promoted to executive vice president of the parent company. R. C. Dixon succeeded him as Anadarko president and continued O'Shields' strategy.

The friendly and easygoing Dixon decentralized Anadarko operations, establishing regional offices in Houston, Denver and Calgary, along with division offices in Oklahoma City and Liberal. The breadth of locations reveals that, by the end of its first decade in business, Anadarko had expanded into most major onshore exploration areas in the United States. Dixon then turned his attention to the Gulf of Mexico. In 1970, Anadarko participated with five other companies in acquiring drilling rights to nine blocks offshore Louisiana at a federal lease sale.

Bob Allison in 1976

Bob Allison was 37 years old when he took the helm of Anadarko as president in 1976.

Two years later, it joined a second group of companies in leasing 13 blocks offshore Louisiana, and yet another group that acquired a 5,000-acre block 10 miles from the Trunkline system. Panhandle leveraged its burgeoning offshore assets in 1974, building the Stingray gathering system, which at the time was the largest offshore gas transmission system developed as a single project in North America.

Although Anadarko was not the operator on these early offshore plays, "we were beginning to get our feet wet, learning as we went," says Robert J. "Bob" Allison Jr., who was hired in 1973 as Anadarko's vice president of operations.

"We still had to develop a staff of skilled offshore personnel and the expertise to make sure we were handling our investments properly."

Anadarko's first production from the Gulf of Mexico flowed into the Stingray system from one of two platforms installed in 1974 at Vermilion block 320, producing gas at a rate of 100 million cubic feet per day. "Vermilion, in which we had about a 20 percent interest, didn't provide a huge boost to production, but it was important psychologically and morale-wise that our work in the Gulf held great potential," Allison says.

Dixon's successor as Anadarko president was Robert M. Stephens, a former Tenneco executive appointed to the post in 1971. Stephens restructured and downsized Anadarko's organization, which had ballooned after the Ambassador acquisition and was in need of pruning. "There was a lot of dead wood in the company," Gary Cobb confides. "Stephens did what was needed, and it got us going in the right direction." Stephens also moved Anadarko's headquarters in 1974 to the fifth floor of a modern, 12-story office building on Allen Parkway in Houston.

Over a period of years in the 1970s, the restructuring gave Anadarko ownership of significant gas holdings that Panhandle had kept when it originally formed Anadarko. For the same reasons of price inequity that had led to Anadarko's birth, Panhandle created two sister subsidiaries, Pan Eastern and Pan Western Exploration. With the FPC's approval, Panhandle transferred its remaining gas-producing properties into the new units. In return, says O'Shields, Panhandle "had to invest the additional revenue we received into an exploration program focused on finding natural gas, which would help ease the gas shortage."

Pan Eastern Exploration soon mounted several ventures to explore areas near Panhandle's legacy transmission systems and later branched into offshore exploration in the Gulf of Mexico. By 1976, Pan Eastern and Pan Western — since renamed APX and APX Western — essentially had been rolled into Anadarko. Pan Eastern alone had given Anadarko an estimated 1.3 trillion cubic feet of gas reserves.

Hugoton Field Sunset

Anadarko Production Company, as the company was called prior to its independence, has historic ties to the natural gas-rich Hugoton field, stretching from the Dust Bowl of southwestern Kansas into Oklahoma. Panhandle Eastern founder Frank Parish originally bought long-term leases from landowners in the vast field at an average of $10 an acre.

Bias Toward Exploration

As the United States celebrated its bicentennial in July 1976, Anadarko employees toasted a new president — Bob Allison. The company, numbering 292 employees, was on the launch pad of becoming a major independent exploration and production company. Allison lit the fuse that sent it soaring. Under his leadership, Anadarko began

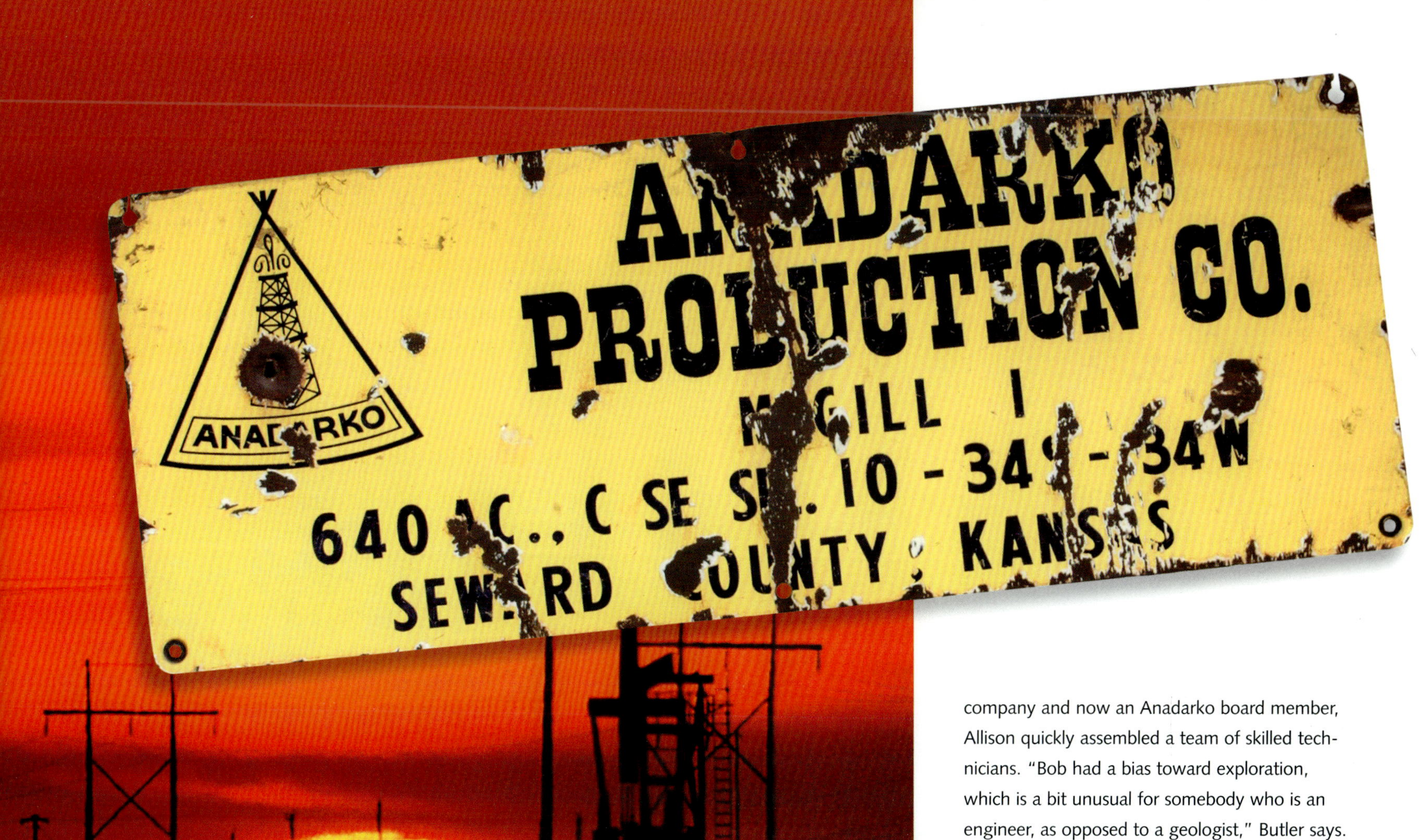

to take the lead as an oil and gas platform operator, developing its own geological and geophysical data to justify acquiring and drilling offshore. "Our strategy was to make money by finding and producing oil and gas, not just as an acquirer or a developer but as a full-service company with exploration as the key," Allison says. "This required skilled technical people, a strong lease position in promising areas and cash flow to sustain an aggressive exploration program."

Anadarko already had accumulated a substantial lease position in the major U.S. onshore oil and gas basins, such as the Hugoton field, and was building an impressive presence offshore Gulf of Mexico. According to John R. Butler Jr., at the time a reservoir engineer at a competing company and now an Anadarko board member, Allison quickly assembled a team of skilled technicians. "Bob had a bias toward exploration, which is a bit unusual for somebody who is an engineer, as opposed to a geologist," Butler says. "But he invested in state-of-the-art geophysical technologies to give his explorationists the tools they needed."

Meanwhile, James T. "Jimmy" Rodgers had filled the vice president's position that Allison had vacated. Rodgers brought additional financial discipline to exploration, subjecting a prospect property to rigorous technical and economic analyses before it could be drilled. Having seen a number of companies in the industry go bankrupt during pricing upheavals, he decided Anadarko needed a new standard by which to determine how much to invest in a property — one that would be reliable even when market prices were unstable.

"We were pretty traditional at the time in

the oil industry, making investment decisions based on the usual economic criteria like rate of return and return on investment," Rodgers says. "These metrics rely exclusively on a company's future forecast for oil and gas prices. If you want to drill a well today, you have to forecast the future cash flow from the sale of products from that well."

Anadarko became one of the first companies to use a new economic standard called "cost of finding" — a measure of exploration and development costs per barrel, which gave a quick way to determine the economic promise of a prospect or project. "If drilling costs are high on a prospect and the potential reserves are relatively small, the cost-of-finding criterion eliminates that prospect," explains Rex Alman, former Anadarko senior vice president of onshore operations. "We used cost of finding in addition to the other tests common in the industry to ensure we were focused on the right plays."

The reinvigorated exploration program bore fruit in 1979 with two finds in the Gulf of Mexico. The discoveries at the West Delta 138 and High Island A-376 blocks marked the first and second times Anadarko took the operator position on an oil and gas production platform. Although West Delta proved a commercial disappointment, High Island earned money and was extolled throughout the industry for its technical achievements. "We set a stationary eight-leg platform in water that was 327 feet deep, a depth record in the Gulf at the time," recalls Chuck Abernathy, the Anadarko engineer whose economic models helped justify the building of the platform and who later became vice president of operations. "We drilled close to 36 wells on that thing." When the High Island platform came online in 1983, initial production was 60 million cubic feet of gas and 8,000 barrels of oil per day. Gas production from High Island alone represented almost a third of Anadarko's daily production at the time.

Can't Find a Rig

Ironically, neither West Delta nor High Island was the first Gulf of Mexico block operated by Anadarko to begin production. That distinction belongs to another Anadarko-operated platform at the East Cameron 104 block offshore Louisiana.

West Delta, East Cameron and High Island were the opening scenes to Anadarko's climactic play in the Gulf of Mexico — Matagorda Island field off the Texas coast, the largest offshore natural gas discovery in the company's history at the time. In 1979, Anadarko contacted Amoco Production Company, whose then-president believed, as did many in the oil patch, that there was little or no natural gas offshore Texas. Anadarko arranged to farm-in, agreeing to pay all of the drilling costs on the first well in exchange for a one-half working interest on the Matagorda Island 623 block, to which Amoco held the lease — clearing Anadarko to explore and drill there. Time was of the essence, however. Amoco's lease on block 623 was set to expire in early 1980, and offshore drilling rigs were scarce due to a boom in energy exploration. "When we went looking for a rig to drill block 623 in October 1979, we couldn't find one," Allison recalls.

The company located a drilling ship, but Allison was adamant that Anadarko not use a floating rig in 80 feet of water. It turned out to be a prescient decision. "We eventually found a rig that was loaned to us by another producer," Allison explains. "While drilling the well, we had a blowout. Had we had a blowout with a floater, it could have been disastrous for everyone aboard the rig."

The first well drilled at Matagorda Island 623 in late 1979 indicated a modest discovery. Since the farm-in deal allowed Anadarko to operate the project until a discovery was made, Amoco had the right to get back in as the operator with a 50 percent working interest in the block, which it proceeded to do. Anadarko previously had sold 25 percent of its position to Champlin Petroleum Company, leaving Anadarko with a 37.5 percent working interest.

"An Explorer's Dream"

As the partners prepared to install a platform at the 623 block, Robert Talley, Anadarko's director of geoscience technology who was then a senior Anadarko geophysicist, undertook seismic studies of the surrounding area. The data indicated the presence of a promising structure on the 623 block's western edge. The companies drilled a second successful well, and additional geophysical mapping revealed evidence of yet another structure, this time in the adjacent block 622. As the deadline on a federal lease sale that included the 622 block fast approached, the partners drilled a third well on 623 to evaluate the adjacent block. What they saw gave them the confidence to bid $43 million on the lease for block 622, which they won despite high industry interest.

With fingers crossed, the group drilled its first well on block 622. It encountered more than 400 feet of pay — a major discovery. Along with the wells on block 623, the partners had found a vast reservoir of gas. "It was an explorer's dream," says Talley. "I remember the exploration manager on the project turning to me and saying, 'Robert, you'll probably never see another discovery like this in your lifetime.'"

First from the Gulf

Anadarko's first production from the Gulf of Mexico originated at the Vermilion Block 320 offshore Louisiana. In the mid-1970s, the field produced gas at the rate of 100 million cubic feet per day.

The discoveries on the Matagorda Island blocks were especially rewarding since all of them were "wildcats," wells drilled in unexplored ground. Anadarko's technical expertise and scientific rigor had won the day. "Even though Amoco was the operator, we were very influential in getting our ideas implemented," Alman says. "Matagorda gas is deep, hot and over-pressured. It took a lot of very unique procedures to optimally complete these wells — a lot of good science, which we had in spades."

One of the earliest 3-D surveys undertaken in the Gulf of Mexico by the partners provided data on how best to delineate and drill additional wells around each of the discoveries. "We ended up needing three platforms plus several individual well tiebacks," recalls Talley. "It's amazing when you think we had fewer than 40 geophysicists on the payroll then. We were a very small company and yet we had helped discover the largest gas field ever found offshore Texas — more than a trillion cubic feet of natural gas." Production from the field eventually came online in 1985-86. By year-end 1998, the field had produced 899 billion cubic feet of gas and 9.3 million barrels of oil.

While Matagorda was a home run for Anadarko, there were plenty of base hits that also added to production. The company became proficient at enhanced oil recovery in the 1980s, employing various methods to squeeze oil from older properties that seemed to have offered their last drops. So-called secondary techniques included waterflooding — injecting water into existing wells to boost the natural reservoir pressure, forcing oil toward producing wells and increasing the total amount of oil recovered. Tertiary practices used caustic chemicals, polymers or carbon dioxide to achieve the same end.

Anadarko's first tertiary recovery project was in the Interstate field in Morton County, Kansas, in 1982; it ultimately produced an additional 1.6 million barrels of oil. That same year, the company operated 32 waterflood properties, which provided 26 percent of its total oil production. Such unconventional plays soon became a core competency.

Four hundred employees worked at Anadarko as the 1980s began. Bob Allison and his disciplined management team had ushered the company out from under Panhandle's shadow, chalking up new discoveries and reserve replacements that earned the industry's respect and brought the company renown.

Allison had his sights on further growth, and to achieve it he needed top-notch geologists, geophysicists and engineers. In 1980, he relocated Anadarko's headquarters to Greenspoint in North Houston, close to the airport and the homes of a preponderance of oil-industry workers. He then launched a recruiting effort, offering prospective employees a chance to join a fast-growing, nimble and secure organization that put a premium on collaboration. Bruce Stover, chief engineer at the time who eventually became Anadarko's vice president of worldwide business development, started a program to recruit engineers and geoscientists directly from college — an innovative approach for such a small company.

"We truly had an integrated effort," Alman says. "Unlike many larger companies, where the exploration people were on one floor and the production people on another and no one talked, we were right next door to each other and talked all the time."

Allison earmarked more than $228 million for exploration and development in 1980, three times the amount invested in 1979. Anadarko was active in most of the major geologic areas in the United States, wildcatting in promising regions like the Cannonball well in Montana and North Dakota, which earned the company its first mention in the *Wall Street Journal*. The bulk of the company's budget, however, was directed toward known petroleum provinces like Kansas and Oklahoma, with a smaller percentage geared to overseas exploration, where the risks were greater but so was the potential for bigger reserve additions. Anadarko's budding international exploration focused on interests offshore Bahrain, offshore Italy and offshore Sri Lanka, but these efforts were unsuccessful.

Most of the company's total reserves were natural gas. "In the early eighties, we were about 95 percent gas and the rest crude oil," Allison says. "As we started to make presentations to Wall Street separate from Panhandle, we were seen as unbalanced — as purely a natural gas play. It was felt by Wall Street that it would be smart for us to have a little more oil as a buffer to volatile gas prices."

A more balanced Anadarko emerged in 1984, the subsidiary's 25th anniversary year. Crude oil production reached 4.9 million barrels (about 13,400 barrels a day), a company record at the time. Anadarko's lease holdings now exceeded two million acres in the principal producing areas of the United States, and a growing lease and production position in Western Canada. In just a quarter of a century, the small enterprise in Liberal, Kansas, whose 50 employees split their work between two companies, had become one of the five largest non-integrated companies in the oil and gas industry, in terms of proved domestic reserves. It

Seismic Shoot

Robert Talley, manager of geoscience technology, and Danny Addis, a geophysical advisor, oversaw seismic studies of Matagorda Island in the late 1970s that helped lead Anadarko to a gas reservoir so large that Talley calls it "an explorer's dream." Above, Talley, at far right, and Addis, to Talley's left, prepare to board a helicopter with crew members en route to a seismic shoot.

At Home in Kansas

Anadarko has an especially deep connection to Kansas, shown here at sunset. It is where its predecessor, the Missouri-Kansas Company, got its start in 1928. When Anadarko became a subsidiary of Mo-Kan successor Panhandle Eastern Pipe Line Company in 1959, its first headquarters was in Liberal, Kansas. The Hugoton Embayment in southwestern Kansas and Oklahoma contributed much of the company's early production.

also was considered the leading non-integrated energy company, as measured by the volume of reserves added per dollar of investment.

There was little question that Anadarko not only was the standout performer among Panhandle's profit centers. It was the company's crown jewel. Yet the parent was painfully aware that shareholders weren't reaping Anadarko's full value in the price of Panhandle Eastern stock, which traded on par with other diversified pipeline companies — discounted by a third of the actual value. As then-Panhandle CEO O'Shields puts it, "The whole of Panhandle was not as valuable as the sum of the parts, particularly with respect to one part — Anadarko."

Raiders on the Prowl

Panhandle had a plan to change the status quo. It restructured into a holding company with profit centers like Anadarko set up as wholly owned subsidiaries. Anadarko Production Company was renamed Anadarko Petroleum Company, and it began the practice of breaking out operating results in its own annual reports. The message was clear — Panhandle was preparing to spin off Anadarko for the benefit of stockholders. But a thicket of problems, including burdensome regulations, an unfriendly takeover bid and a costly contract dispute with a major supplier, stood in the way.

The first obstacle actually had been put in place several years earlier when Congress passed the Natural Gas Policy Act of 1978, a response to industry-wide gas shortages so acute in 1976 and 1977 that some schools and factories were forced to close. With the artificial price ceilings still in effect for pipeline companies, selling prices for gas often were below finding costs; therefore the companies still had little incentive to add to reserves. Then, in the late 1970s, demand skyrocketed as supplies dwindled. The 1978 gas act sought to cure the industry's decades-long exploration inertia by mandating assorted pricing tiers. It created 28 separate categories of gas, each with a separate price depending on whether the gas was old or new, deep or shallow, onshore or offshore, or produced by a large producer or a small one.

The legislation initially seemed to work. Gas prices rose, exploration boomed and pipeline companies increased their reserves, with several engaging in long-term contracts to supply natural gas — but demand dropped just as the shortages were easing. Gas consumption fell from 19.4 trillion cubic feet in 1981 to 16.8 trillion cubic feet in 1983. Producers were left with a supply glut, dubbed the "gas bubble," in 1982 just as prices tumbled. Several pipeline companies, Panhandle among them, had locked into long-term, take-or-pay contracts with suppliers when prices were high, meaning they still had to pay the agreed-upon prices even if they didn't take the gas.

For its part, Panhandle had signed a 20-year agreement with Algeria's state-run oil company, Sonatrach, to purchase and import 3.3 trillion cubic feet of liquefied natural gas (LNG) from Algeria, beginning in 1981. The deal further required Panhandle to invest in three tankers to ship the gas as a liquid, and to build a receiving and storage facility in Lake Charles, Louisiana, to re-gasify it for Midwestern markets. Panhandle had earmarked $1 billion for the Lake Charles plant alone.

The contract made economic sense when gas supplies were short and prices were high — but not when the shortage turned into a surplus and gas prices fell precipitously, putting Panhandle in the position of having to buy gas it could only sell at a loss. Its survival threatened, Panhandle refused to buy the agreed-upon volume of LNG. Sonatrach viewed this as a breach of contract and filed for arbitration in Geneva, Switzerland.

In the meantime, it became clear that Panhandle was not alone in recognizing the unrealized value of its Anadarko subsidiary. As the company sought a way to position Anadarko's assets for shareholders' benefit, such as selling the company or spinning it off to stockholders, an unwelcome suitor arrived. In June 1986, oil and gas producer Wagner & Brown, Ltd. of Midland, Texas, launched a hostile takeover offer for Panhandle at double the company's stock price. Panhandle rejected the offer, believing it too low. Nevertheless, a Sword of Damocles hung over the old pipeline company. As Paul Taylor, then Panhandle's investor relations director, recalls, "There were plenty of Wall Street raiders who were prowling, looking for the right way to cut Panhandle into valuable pieces."

The takeover attempt intensified Panhandle's efforts to settle the dispute with Sonatrach, clearing a path to spin off Anadarko to shareholders. In the summer of 1986, sleepless executives of both companies met in Geneva, Paris and other European cities to craft an amicable solution.
In July 1986, they finally breached the impasse: Panhandle agreed to pay Sonatrach $300 million and give it six million shares of Panhandle Eastern stock, a roughly 10 percent stake in the company, to settle the LNG debacle.

Two months later, Panhandle spun Anadarko off to shareholders as a separate, independent company. Stockholders received one share of Anadarko common stock for each share of Panhandle common stock they owned. In the words of Allison, "We were on our own now."

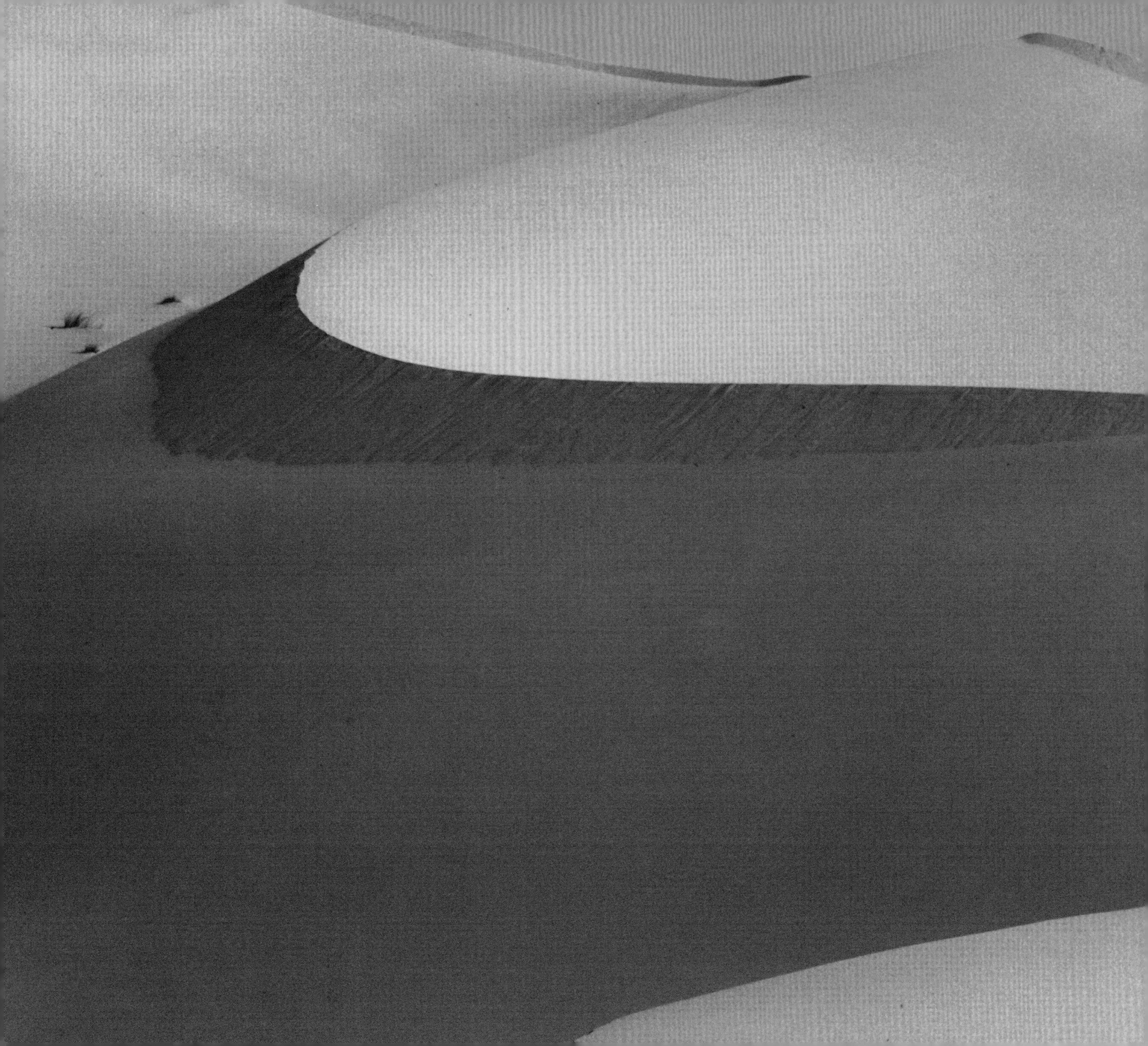

CHAPTER THREE: BREAKING OUT

Independence Day

On October 1, 1986, Anadarko Petroleum Corporation officially became an independent company. At the New York Stock Exchange that day, Anadarko Chairman and CEO Bob Allison purchased the first 100 shares of common stock. Allison, left, and Anadarko Chief Financial Officer Mike Rose, right, witnessed an NYSE staff member recording the sale.

On September 8, 1986, Anadarko shares began trading on the New York Stock Exchange. Three weeks later, on October 1, the spinoff of Anadarko was made official, and Chairman and CEO Bob Allison and Chief Financial Officer Mike Rose went to New York to ring the opening bell at the NYSE. Allison stepped up and bought the first 100 shares of Anadarko stock traded that day, at $19 per share.

On the day it became public, the company listed 640 employees and activities in 20 states and three Canadian provinces. Anadarko's former parent, Panhandle Eastern, lived on for a few more years, acquiring Texas Eastern Corporation and a new name, PanEnergy, in 1989, before being acquired itself by Duke Power Company in 1996. Before its demise, Panhandle would wrangle with Anadarko in several controversial disputes. At the moment, though, Allison breathed contentedly.

Allison and Paul Taylor, Anadarko's tireless vice president of investor relations, assisted by Mike Rose, the company's chief financial officer, had mounted an exciting and highly successful road show as part of the initial public offering of stock. Within a year, both Anadarko and Panhandle shares were trading in the $30 range, compared to Panhandle's pre-spinoff trading range of $30 to $40 per share — proof that the spinoff from Panhandle certainly was in stockholders' best interests. Anadarko traded on par with its peer group of independent exploration and production companies, while Panhandle continued to trade with other diversified pipeline companies at a lower multiple.

Poet Ralph Waldo Emerson wrote that an

The Steady Hand

Bob Allison served with distinction as Anadarko's visionary leader from 1976 to 2003, with a year off for good measure in between. Over this 27-year period, Allison adeptly guided Anadarko, navigating rough shoals always with an eye on better seas ahead. To many employees and retirees, he embodies the spirit and success of Anadarko.

Robert J. Allison Jr., was born on Jan. 29, 1939, in Evanston, Illinois, to Robert and Mary Allison, parents who believed in the merits of education in building a boy's character. The family lived humbly in a walk-up apartment across the street from Evanston High School, where Bob loved to play "pilot" as a boy on a World War II U.S. Navy Hellcat fighter plane at rest in the schoolyard. Some of his favorite escapades were as a Boy Scout. In letters written home to his parents, the 11-year-old brimmed with confidence. "I have passed the highest rank in swimming and have taken my canoe test," Bob wrote on July 7, 1950.

Allison attended New Trier and Glenbrook High Schools and later studied for his degree in petroleum engineering at the University of Kansas. To support himself at college, he worked as a butcher, warehouseman, pilot plant operator, roughneck and draftsman. Graduating in 1960, he went to work for Amoco in Duncan, Oklahoma, and later moved to Vivian, Louisiana. On June 17, 1961, he married Carolyn Jean Grother.

The young couple moved to Ulysses, Kansas, where their first child, Amy Elizabeth, was born. The family relocated to Liberal, Kansas, to await the arrival of another daughter, Ann Mary, in 1963. Three weeks hence, the family moved to Jackson, Mississippi, and the following year to Oklahoma City, where another daughter, Jane Susan, was born. The Allison family was always on the move, even spending years abroad in Trinidad, Iran and other countries. "Whenever Bob came home and picked up the globe of the world, his daughters knew they were moving again," says Mary Stehling, Allison's longtime senior executive assistant.

In a letter, Bob's and Carolyn's youngest daughter Jane listed the various places she lived while growing up, concluding, "In 1973, we moved to Fort Worth where Dad went to work for Anadarko as vice president of operations, and he finally got a house on a golf course that had a PGA tournament. In 1974, we moved for the last time to Houston."

In his early days at Anadarko, Allison was a man in a hurry, says Ellen Cobb, who worked at the company for 48 years. "Bob was always running up and down the halls at our Fort Worth location," Cobb recalls. "He was so friendly and open, the kind of person who never forgets your name, just as he is today. To many of us, he *is* Anadarko."

Bob Allison displays a seismic image of the Mahogany subsalt field. Advances in 3-D seismic technology allowed Anadarko to see through the salt to a million-barrel field below.

Big HUGS

The Hugoton Gathering System connected 138 producing wells, giving Anadarko greater control of its natural gas assets. HUGS was designed by Mike Ross, at left, Anadarko general manager, midstream operations. Also influential in Anadarko's onshore success was Jay Smith, at right, former U.S. onshore manager.

"institution is the lengthened shadow of one man." In Anadarko's case, this man is Bob Allison. An Illinois native, Allison evidenced outstanding leadership as a boy. At the age of 11, he was elected patrol leader of his Boy Scout troop, a "trying job" requiring him to maintain order despite his charges' tendency for "horsing around," he complained to his parents in a letter. Allison kept them in line and, before summer's end, was promoted to Second Class Scout, a rare achievement for a first-year Boy Scout.

Allison studied petroleum engineering at the University of Kansas and joined Amoco shortly after graduating in 1960. In 1973, he joined Anadarko, where, as vice president of operations, he caught the eye of Panhandle executives such as then-Chairman and CEO Dick O'Shields. "Bob was charismatic and razor sharp," O'Shields says. "He was extremely intelligent but also broad-minded. It's why I made him president of Anadarko."

Allison was named president of Anadarko in 1976 and became chairman and CEO when it went public. He worked on an entirely different level from his peers in the exploration and production business — in his grasp of the big picture, his instinctive feeling for the right plays and his telegraphic gift for judging a person's aptitude and qualities. He had an open-minded approach to exploration, rare for an engineer. "Bob brought an engineering discipline to exploration that gave this company balance," says Charles Manley, former executive vice president, administration. "When you layer on his business acumen and financial discipline, it gives you a good feel for Anadarko's success, but he's the last guy to wear his intellect on his sleeve. Bob doesn't have the need to impress anyone."

Allison assembled a highly skilled team of geologists, geophysicists and engineers — "the best in the business," he often said. He never expected to become expert in every phase of exploration and production, trusting instead the expertise of his staff. "The big thing with Bob was trust — if you gained it, you had tremendous latitude and freedom to run your area, and he would back you to the end," says Manley. "He maintained an overview of your work and had some contact, but he was never a micro-manager."

Robert Talley, manager of geoscience technology, found it liberating to work under Allison. "He was just the easiest person to present (a project) to," Talley says. "He would ask incisive questions about the particular play or the technology, but he always let you run with the ball. The trust and support we receive in our work is one of the core strengths of Anadarko."

Allison demanded performance but did not expect short-term results. He understood that it can take years to develop a prospect, gain a land concession, undertake the technical work and drill the wells — possibly all for naught. "To be successful as an explorationist, Bob knew he had to build a team of great technical professionals and then keep them," Manley says. "He did that by assuring them that they shouldn't be afraid to drill a dry hole — as long as they did their homework, Anadarko would learn from it."

Sticking with Employees in a Down Market

Launching the second stage of its history as a fully independent company, Anadarko was blessed not only with insightful management, skilled scientists and great technological tools but also with an estimable portfolio of assets. The company's proved reserves in 1986 of 313 million energy equivalent barrels (a metric combining crude oil and natural gas reserves) ranked it as one of the nation's largest independent exploration and production companies. Its reserve replacement in 1986 nearly doubled 1985 production, at one of the lowest finding costs in the industry.

On the down side, the company couldn't have picked a more difficult oil and gas market in which to make its debut. After slipping for three years, crude oil prices collapsed in 1986, and natural gas prices weakened from the persistent imbalance in supply and demand. The lower prices and volumes for oil and gas took a toll on Anadarko's bottom line, causing revenue and earnings to plummet. "Here we were this brand new, independent NYSE-listed company, and it was a horrible time in the business to make any money," Allison remembers.

Undaunted, the CEO was confident that Anadarko's intellectual capital more than made up for the loss in financial capital. Unlike his peers who downsized their companies when markets turned south, Allison believed in retaining employees to advantage the inevitable upturn. "Bob didn't want to lose the company's institutional knowledge and expertise during a downturn," explains John Butler, a member of Anadarko's board of directors. "He refused to decimate the technical capabilities of this company, always insistent that it was people who made an organization go forward."

"Cost of Finding" Promoter

The price of exploration is a critical component in the hunt for oil. To help the company better determine the value of prospects, in the late 1970s Anadarko developed a "cost of finding" metric. Jim Ashton, at the time Anadarko's supervisor of planning and economics, was tasked with promoting the new metric to employees.

Doubling Production

Anadarko's historic assets in the Hugoton field — more than 400 shallow wells — were the company's most important producing properties in its first years of independence. Mark Pease, senior vice president of E&P technology and services, above at left, was Anadarko's division manager in Liberal, Kansas, at the time. He led an infill drilling effort — taking advantage of new regulations permitting the drilling of additional wells, such as the one opposite, to drain the reservoir. The effort doubled Anadarko's production.

Many Anadarko employees recall the market conditions of 1986 and their concerns over job security. "The big oil companies were laying off people left and right, but Bob said he would stick with us, even after he got some heat from Wall Street," says Todd Fowler, a geologist who began his career at Anadarko in 1980. "Those of us with gray hair remember that." Fowler kept his job even though his drilling budget was gutted because of the pressure to contain costs. "Bob told me my budget was reduced to zero; I said to him, 'Are you sure you still need me?'" Fowler adds. "He told me to keep prospecting, to build an inventory of wells that we could drill once some money came in." Six months later, the markets revived and Fowler resumed his drilling program.

When Anadarko was reborn as a public company, its most important producing properties were in the Hugoton field, where it owned interests in more than 400 shallow natural gas wells. Just when it needed it most, the company would gain the ability to augment production from the basin. At the time, most of the original wells drilled in the field in the 1930s and 1940s produced from only the top three of five gas-bearing zones, leaving potentially recoverable gas in the two bottom zones. Existing regulations limited producers to only one well per 640 acres, yet reservoir engineers were convinced that a single well could not efficiently drain all the reserves.

Prospects brightened when the Kansas Corporation Committee (KCC), the state regulator charged with ensuring that the Hugoton reservoir was depleted efficiently and equitably, held public hearings on expanding the drilling options. The KCC mulled allowing "infill" drilling — the drilling of a second well in each 640 acre-plot to produce the remaining reserves. Producers like Anadarko strongly favored infill drilling — in part because the Federal Energy Regulatory Commission let them charge higher prices for gas produced from infill wells — but the proposed rule change encountered stiff opposition from interstate pipeline companies like Panhandle. They argued that the producers already had more than enough gas to sell from the Hugoton field.

In 1986, the KCC sided with the producers, asserting that the Natural Gas Policy Act of 1978 expressly required producers to obtain the most complete production of natural gas possible from domestic reservoirs. After the KCC rule went into effect, infill gas was priced roughly four times higher than legacy gas. The timing for Anadarko was serendipitous.

"Oil prices had collapsed in 1986, compelling a lot of exploration and production companies to

reduce their activities; meanwhile, our activity actually ramped up thanks to infill drilling for gas," explains Mark Pease, Anadarko's division manager in the Liberal office at the time. "We added second wells in many of our 640-acre plots, more than doubling our production." Anadarko drilled more than 260 infill wells in the basin. When prices rose, the company was poised to reap the benefit financially.

Challenging the Mothership

As Anadarko took flight as an independent public company, it often ruffled feathers at Panhandle. In negotiating the spinoff, Anadarko had agreed to supply natural gas to Panhandle at prices that Allison believed were not in the best interest of Anadarko shareholders. Taking a tough stance against his former employer, he argued that agreements made between the companies were unfair because both corporations had common board members at the time they were signed. "Bob's attitude was that we didn't have to agree to the onerous terms anymore and could renegotiate them," Paul Taylor recalls. "Panhandle said they wouldn't renegotiate. Consequently, we rescinded the agreements. When that didn't resolve the dispute, we sued them."

The case ultimately landed in the Chancery Court of Delaware, a state court specializing in resolving corporate conflicts (both companies were incorporated in Delaware). O'Shields, Panhandle's CEO, was confident of winning. "The lawsuit was improvident," he says. "The premise was that we had forced Anadarko to enter into contracts with us that they would not have entered into otherwise, had they already been independent. They argued that the contracts hurt shareholders, which was ridiculous since we both had the same set of shareholders."

Panhandle won the case. Nevertheless, Taylor believes the clash between the companies was necessary. "It showed that Anadarko was not afraid to flex its wings and test the limits of its independence," he says. "Panhandle thought they were getting rid of this wonderful stepchild that would go out on its own and conquer worlds and never forget where it came from. But Anadarko had a CEO named Bob Allison who was determined to be totally independent — even if it meant suing the parent company."

Anadarko lost the battle but eventually won the war. A ruling by the Federal Energy Regulatory Commission in 1986 paved the way for producers to renegotiate certain agreements they had in place with pipeline companies. In its Order 451, the FERC stated that price controls on "old gas" — natural gas committed to interstate commerce when the Natural Gas Policy Act of 1978 was enacted — were unjust, because they were below replacement cost. The ruling collapsed the act's many price tiers into one, with a ceiling price set above the market-clearing price. Producers rejoiced; pipeline companies bellyached.

Armed with Order 451, Anadarko approached Panhandle to renegotiate its supply contracts. Panhandle rejected the request, arguing that it would terminate the contracts before it would agree to a higher price. "Allison said, 'Let them cancel. We don't need them,'" Taylor recalls. "I said, 'Wait a minute. We're going to cancel our biggest contract, representing more than 70 percent of our revenue?'

Discovery in the Gulf

Anadarko recorded its first major Gulf of Mexico discovery at Matagorda Island Block 622/623, with production commencing in 1984. As production engineers in the 1980s, Danny Rea, current Anadarko vice president, midstream, conferred with Steve Pearson, current vice president, operations – eastern region, at Matagorda Island.

Technology to the Rescue

Anadarko is widely respected in the oil and gas industry for its revolutionary approaches to difficult plays. From its earliest years as a subsidiary of Panhandle Eastern, the company has put its faith in its skilled geologists, geophysicists and engineers to devise novel approaches to coaxing oil and gas from recalcitrant fields — drilling for oil where other companies have not dared to tread or have given up, and taking large risks where the payback warranted.

"We've had a good track record in the industry over the past 20 years of finding giant oil and gas fields with more than 100 million barrels of oil equivalent, and a lot of it was technology-driven," says Bob Daniels, senior vice president of worldwide exploration.

In the early 1990s, Anadarko became a leader in hydraulic fracturing, a set of processes in which fluids are pumped at a high rate into a reservoir, cracking the rocks to provide a path for natural gas to rise to the surface. The technology was successfully applied at the company's Golden Trend play in Oklahoma, a field forsaken by the industry. The company also revived the Bossier Trend in Texas, which much of the industry, including Anadarko, had abandoned in the belief that it was no longer commercial. Anadarko returned to the scene with more durable and efficient PDC (polycrystalline diamond compact) drill bits and a hydraulic fracturing concept that pumped water into the rock instead of more expensive gel-and-sand mixtures. These cost-effective technologies were later applied successfully in Louisiana's Vernon field and other tight gas plays.

Offshore, Anadarko's technical skill was evident at the High Island A-376 platform in the Gulf of Mexico. In 1990, the company drilled wells from the platform using a large, semi-submersible drilling rig anchored alongside as a tender to eliminate the need for a derrick barge — an industry first. The offshore arena also was the site of Anadarko's groundbreaking efforts in the Gulf's subsalt trend. For decades, explorationists had postulated that large hydrocarbon structures were buried beneath the 300-mile wide dome. A pioneer in the use of 3-D seismic technology, Anadarko wielded advances in this technology to obtain the first reliable glimpse of rock formations beneath the dense salt formation, launching a new round of Gulf exploration.

In 1990, Anadarko drilled wells from the High Island A-376 platform using a semi-submersible drilling rig anchored alongside as a tender — an industry first.

Today, the company's technological prowess is seen in the Gulf of Mexico's deepest waters. In the Green Canyon, about 160 miles south of New Orleans, Anadarko discovered the Marco Polo field in 4,300 feet of water — a record in 2000. With its partners, the company built the deepest tension-leg platform in the world at the time, which is tied to the Gulf floor by eight 28-inch-thick steel cables. In the untapped Eastern Gulf of Mexico, Anadarko is involved in ultra-deepwater exploration in water depths of more than 9,000 feet. The company also is developing Independence Hub, a floating platform capable of producing an estimated one billion cubic feet of natural gas per day when production commences in mid-2007.

The Marco Polo field was Anadarko's first deepwater find in the Gulf of Mexico. Its hydraulic umbilicals, pictured opposite top, control the production rate and safety valves on the Marco Polo wellhead. Independence Hub, left, a floating platform in 8,000 feet of water in the Eastern Gulf of Mexico, was expected to set a record for the deepest offshore producing platform in the world when installed in late 2006 or early 2007.

And Bob said, 'I don't care if we lose our biggest customer. I'm just not going to give the gas away. If they don't want it, somebody else will.'" Rather than wait for Panhandle to act, Anadarko itself cancelled the contracts — and as Allison had predicted, the company had no trouble finding other buyers at higher prices.

Order 451 was one of the first sallies toward eventual deregulation of the natural gas markets. In 1989, the federal government deregulated natural gas producer sale prices. Four years later, the market completely determined the price of natural gas.

Only a year after Anadarko joined the NYSE, *Business Week* ranked it as the 406th largest company in the nation, based on total stock market value. *Oil and Gas Journal* listed Anadarko 19th in the United States and 20th in the world in natural gas reserves. "We've got the best track record in the business (for) finding oil and gas profitably," Allison told the *Houston Post* in 1987. He was not off the mark. Half of the 48 exploration wells drilled in 1987 were discoveries. One of the more significant finds was at Matagorda Island block 587, a field touted to produce 100 billion cubic feet of gas.

The energy markets finally revived in 1987. Allison's no-layoff policy during the bleak times paid off: proved energy reserves in 1987 were double what they had been five years earlier, positioned to be sold at much higher prices. The company had grown at the drill bit rather than through acquisitions, the tactic adopted by many competitors. As business progressed, Anadarko's revenue increased 45 percent from 1987 to 1988 to $333.1 million, and its earnings quadrupled.

Anadarko spent most of the windfall on exploration and development. It also invested more than $23.3 million in acquiring oil and gas producing properties, many located in the Permian Basin in West Texas, where it had established a regional office in Midland in 1978. Many properties were no longer considered economically viable by their previous owners, and Anadarko implemented secondary and tertiary recovery techniques to squeeze more oil from the reservoirs. "We started waterflooding them, and by the time we were finished there was no low-hanging fruit left," says Steve Martin, then an Anadarko engineer.

Anadarko also invested in the Arkoma Basin in southeastern Oklahoma and western Arkansas, one of the nation's most important natural gas plays in the late 1980s. The company acquired 29,000 net lease acres in the Arkoma Basin in 1989, primarily to evaluate pay horizons in the deep Arbuckle and thrusted Spiro formations. Ultimately, it participated in 45 wells, drilling in depths from 10,000 feet to 18,000 feet, achieving an impressive 60 percent success rate.

These acquisitions were a mere warm-up. In 1992, Anadarko made the largest onshore acquisition in its history in the Permian Basin, investing $190 million in oil and gas producing properties with total estimated reserves of 34.2 million barrels of crude oil and 60.6 billion cubic feet of natural gas. Although a large number of the fields were mature, they were far from depleted. More than 2,700 wells were still active — ripe for the company's enhanced oil recovery methods. Anadarko ultimately yielded an excellent rate of return on these plays.

As the 1980s drew to a close, Allison's overarching strategy remained the same — balanced exploration in North America and overseas. Anadarko's growth relied on an aggressive search for oil and gas, whether market conditions were good or bad. Rex Alman, former Anadarko head of U.S. onshore operations, credits John Seitz, then general manager of exploration; Mike Cochran, at the time chief geophysicist; and Lee Petersen, chief geologist; with "taking exploration from a shotgun approach to a rifle approach" during the 1980s. According to Petersen, Cochran, who went on to become Anadarko's vice president of exploration, made an especially valuable contribution as the first person to apply "real options analysis" theory to exploration. This improved Anadarko's decisions about whether to develop or abandon exploration assets by viewing them much as an investor would assess the cost and value of stock options — with two price components: a core "strike" or development cost and a separate "premium" cost, which included lease, exploration and appraisal expenses.

As a result of their work, the company focused its exploration plays only on those prospects offering long-term opportunity and developed them with strict adherence to cost of finding, rate of return and other key metrics. Its singular approach to risk management earned it an "A" rating from all three major debt-rating agencies, the only independent E&P company in the United States to achieve this grade.

"We worked hard on achieving a balanced mix of plays in the 1980s with the right blend of

Permian Play

The Permian Basin of West Texas is one of the most prolific basins in North America. Anadarko began exploring in the region in the 1970s and invested $190 million in oil- and gas-producing properties there in the '90s. Since then, the company has used innovative enhanced oil-recovery technologies to squeeze oil and gas from area fields that were mature but far from depleted.

Surrounded by Sand

The sands of the Sahara, home to Algerian nomads and their camels, became the site of Anadarko's greatest exploration success. Despite promising seismic studies of the presence of oil in the sands of Algeria, the risk of failure was high. Several top executives urged CEO Bob Allison to pull back from further exploration after encountering a dry hole. Allison was undeterred, however. His instincts and confidence were rewarded.

risk and reward, drilling in mature provinces as well as unexplored regions," Allison says.

A First in Algeria

Before the decade concluded, Anadarko made news with another major announcement — its first major international exploration deal. In June 1989, the company signed a landmark production-sharing agreement with Sonatrach, making it the first U.S. company to explore Algeria after that country permitted the return of foreign investment into its oil and gas industry. The agreement awarded Anadarko exclusive exploration rights to a 5.1 million-acre concession in the Ghadames and Illizi basins of southeastern Algeria, an area larger than the state of New Jersey.

To manage the risk of exploration in the remote region, Anadarko engaged two partners, LAMSO Oil in the United Kingdom and Maersk, the Danish conglomerate. Each owned a 25 percent interest in Anadarko's share of the project, and together they committed to spend more than $100 million drilling 10 wells over the next 10 years. Optimism ran high, as the Sahara desert was considered one of the most attractive areas in the world to explore for giant oil fields.

Paving the way for the pioneering agreement was the close relationship Allison had formed with Sonatrach officials during the protracted negotiations between Panhandle and Sonatrach in the early 1980s. Allison had earned the respect and admiration of Youcef Yousfi, Sonatrach's CEO, with whom he would negotiate the production-sharing arrangement. "I had known about the geological potential of the Ghadames and Illizi basins, having been privy to technical data produced by a team of Anadarko geologists and geophysicists," Allison recalls. "The technical team and I met in 1986 with Dr. Yousfi in Algiers and told him that we wanted to lease four blocks in the desert, and we were willing to drill 10 wells and invest $100 million. He said he'd never heard of giving away more than one block at a time, and he was surprised at the amount of money we were willing to commit."

Before Anadarko and Sonatrach could finalize a contract, the Algerian government would have to amend its laws to make investment in oil and gas by foreign companies economically attractive. When a new law opening investment doors was passed, Anadarko ended up with almost the exact deal Allison initially had proposed. Reinforcing Sonatrach's interest in the agreement was its 10 percent ownership of Anadarko stock, courtesy of the settlement between Panhandle Eastern and the Algerian oil concern in their 1986 LNG dispute.

The headquarters of Anadarko Algeria opened in Algiers in July 1990. Sixteen Anadarko employees relocated their families and households to the city that summer. They were complemented by Sonatrach geologists and geophysicists along with local drilling and reservoir engineers. In October, this team christened a multi-million-dollar seismic processing center 25 miles east of Algiers and began the exhausting process of collecting seismic data in preparation for drilling the first well. Unfortunately, only a limited amount of recent exploration data was available because the country had been closed to foreign exploration.

Thinking Outside the Box

While team members of Anadarko Algeria coped with unreliable seismic information, their counterparts in Houston were working on a solution. Anadarko was one of the first E&P companies to use 3-D seismic computer workstations, which sped the input of exploration data and the mapping of geologic formations. These efficiencies allowed scientists to spend more time on the vital task of interpreting data, thus improving their success rate in the search for hydrocarbons. "The workstations gave us the ability to move away from paper, helping us be more investigative and thorough in our geophysical understanding of prospects," Robert Talley says.

Anadarko invested heavily in superlative technology and processes — and not just for exploration. The company became a leader in hydraulic fracturing techniques to drain reservoirs that were once considered noncommercial. Fracturing involves the pumping of fluid into a geological formation at such a high rate that the rock cracks vertically, providing an escape route for the gas. Steve Pearson, Anadarko vice president, operations – eastern region, explains, "The best analogy is a crowded parking lot. If you have only one exit, all the cars eventually will get out of the parking lot, but if you have 10 exits they'll get out a lot quicker. Fracture stimulation increases the productivity of a well, thereby improving the rate of return."

Anadarko engineers used hydraulic fracturing in central Oklahoma's Golden Trend, a large field that many oil and gas companies had forsaken. "The Golden Trend play was five reservoirs stacked on top of each other, mostly sandstone and limestone, and we realized over time that it was an unconventional reservoir, with very tight gas sands and low permeability," says Hank

Algeria Field Trip

Brian Sunderland, Allan Driggs, Kevin Lant and Jim Emme, left to right, were members of Anadarko's first Algeria geological field party. The medal, inset, was given to Anadarko by the Algerian government following the signing of the contract with Sonatrach.

DeWitt, former Anadarko senior geological advisor. "Many in the industry felt it wasn't economical anymore, but we saw it as a very viable opportunity — with the right technology. Through creative thinking and tinkering, we made an uneconomic reservoir an economic one."

The company's technical prowess was equally evident offshore. In 1990, Anadarko drilled four wells from the High Island A-376 platform in the Gulf of Mexico using a large, semi-submersible drilling rig anchored alongside as a tender to eliminate the need for a derrick barge — the first time this novel approach was tried.

Such out-of-the-box thinking became a staple at Anadarko. In another example, Steve Martin had a brainstorm in the mid-1980s that resulted in a highly successful oil recovery play. Working in the Liberal office, Martin was searching for a water supply to waterflood the company's Etzold and Hitch oil and gas fields in Seward County, which were producing a scant 100 barrels of oil per day combined. "I knew we could lift the volume substantially through waterflooding, but I couldn't find a good shallow-water zone in that part of southwestern Kansas," he says. "The aquifer was highly restricted."

Martin soon learned that the city of Liberal disposed excess sewage water in the fields beyond the town's sewage pond, where the water evaporated. A light bulb went off. "I thought, 'Why don't we treat that water, pump it up to these reservoirs and inject it?'" Martin says. "We'd even pay the city for the opportunity. It seemed like a perfect win-win scenario."

Liberal town officials jumped at the offer. Anadarko engineers built a water treatment plant in Liberal, then ran a pipeline to transport the treated water to Seward County for injection into the Etzold and Hitch fields. The waterflooding increased oil volumes from 100 barrels of oil per day to 3,900 barrels per day. Anadarko undertook additional injection projects in adjacent fields over the next few years. "Had I proposed this idea to a company other than Anadarko, I probably would have been laughed out of the room," Martin says.

Liberal was the site of another major

Anadarko project, the building of a natural gas gathering system in the Hugoton Basin. The impetus for the project came out of growing frustration with Panhandle — over both the prices Panhandle charged to transmit gas Anadarko produced in the basin and Panhandle's seeming lack of urgency in getting Anadarko gas to market. "When we drilled a new well — and we were drilling quite a number back then in our infill program — it might take a year and a half to connect supply to a pipeline that was only 300 feet away," Allison explains. "I said 'to heck with this' and told Panhandle to consider selling us their Hugoton gathering system. They rejected the offer, so I told them we would simply build our own system on top of theirs. 'You would do that?' they asked. 'In a minute,' I replied."

Anadarko engineers designed a new gathering system and ordered the materials to build it. "I don't think Panhandle really thought we would build the gathering system," says Mark Pease, current senior vice president of E&P technology and services. Allison wasn't bluffing. In May 1991, he smiled as he cut the ribbon at the ceremony

Historic Signing

Bob Allison signed Anadarko's first contract to explore for oil in Algeria following the easing of restrictions in that country. To Allison's left is Mr. Sekfali, deputy director of Sonatrach, Anadarko's partner in the Algerian venture. The pin at left commemorated the long partnership between Anadarko and Sonatrach, while the visa, opposite, enabled many visits to Algeria.

honoring completion of the $28 million Hugoton Gathering System, ironically nicknamed HUGS. The 88-mile system connected 134 producing wells to several pipelines capable of delivering 100 million cubic feet of natural gas per day.

HUGS gave Anadarko the flexibility it needed to move gas at peak demand periods in the winter when prices were high. "It was a huge strategic decision that further defined us as a company that was no longer at the mercy of our former parent," Pease asserts. "Panhandle saw that we were serious and agreed to sell us the rest of their gathering system."

Dry Hole in the Sahara

At company headquarters in Houston, Anadarko was straining at the seams. The Denver office had closed in 1990, and many employees were relocated to Houston. In December, the company began construction of the first Anadarko Tower, a new home in the Greenspoint business complex. The 18-story glass and granite structure with an imposing curved roof was completed in July 1992. Anadarko occupied 250,000 square feet of the tower, taking up most of the floors. As part of the move, the company closed its Oklahoma City office and reduced the staffs at other facilities, relocating these employees to the new complex.

As a page-one article in the Anadarko newsletter *Currents* reported the building of the tower, a smaller item in the same issue outlined tentative plans by Anadarko Algeria to drill its first well in the sands of the Sahara desert. Seismic data had indicated a significant structure in the thick but porous Triassic and Siluro-Devonian sandstone formations.

There were risks, chief among them that the rocks might not have the reservoir characteristics indicated by the seismic data. "No wells had been drilled in this area in years and the potential wasn't really well understood," says Lee Petersen, chief geologist when the original Anadarko Algeria exploration crew was formed. "There also were a lot of people in the investment community that questioned whether we had the ability to drill in the middle of the desert. It was a huge risk and we knew it — a $100 million commitment — but it was something that Bob Allison and John Seitz believed in, so we believed in it, too." Seitz was Anadarko's exploration chief at that time.

On June 20, 1991, the Berkine well, as it was called, was spud. Unfortunately, it was a dry hole. "Instead of millions of barrels of oil, we found an iron mine," Petersen sighs.

Many CEOs would balk at such dire news, but Allison took it in stride. "They told me the well was so dry that I began to visualize dust coming up the through the pipe!" he laughs. "It was the greatest structure you ever saw, but there was no way for oil or gas to migrate to it," says Allison. "But we learned to look for migration pathways."

Clearly, the determined former Boy Scout was not about to let a measly dry hole alter his plans.

Mid-Continent Boost

Anadarko's roots in Kansas stretch back to its earliest days, when the small town of Liberal was the company's headquarters. The Mid-Continent remains an important core of the company's onshore operations, generating natural gas that is piped throughout the region. Anadarko Chief Engineer Steve Martin, opposite as a young engineer in the 1980s, epitomizes the push-the-envelope thinking that has helped turn a 100-barrel-a-day oil field into one producing 3,900 barrels a day.

CHAPTER FOUR: SNOW, SAND AND DEEP WATER

Seismic Shift

Algeria's massive sand dunes stymied initial attempts to find oil in the Sahara desert, as did the severe working conditions. Using new 2-D seismic data, Anadarko reevaluated its geological model after encountering a dry hole. New seismic data gathered by crews, below, guided Anadarko to drill in more opportune locations, and the company found what it had been looking for. A puff of sand, right, from a dynamite detonation signals the firing of a seismic charge.

Few events in the oil and gas business are more unnerving than a dry hole. Not that they are uncommon — most wildcat wells drilled in the 1980s turned up nothing. The case of the Berkine well, however, was perplexing because the available seismic data confirmed the existence of everything required for an active hydrocarbon system — source, trapping, seal and reservoir. The exploration team was certain that a giant oil field lay buried beneath the desert. The question was, *where*?

Anadarko's explorationists were foiled by the Sahara's giant sand dunes, which frustrated their efforts to shoot straight seismic lines. In traditional seismic mapping, geophysicists use dynamite charges to project sound waves into the ground. Measurements of the waves' movement through the Earth tell them the size, shape and depth of the sediment layers below. But in Algeria, giant dunes got in their way, as did the loosely packed sand particles, which produced surface static that masked the seismic response. Existing seismic information was difficult to interpret and, therefore, unreliable.

They were further tested by the working conditions. Access to fax or copy equipment was scant, so the scientists had to use carbon-copy paper in assembling data. Electricity was unpredictable, and clean water for showering was available only between 11 p.m. and 4 a.m. Even the bare necessities of life were wanting. "You couldn't find anything in the shops," recalls Francois Gaulthier, a member of the original Anadarko Algeria team.

Not the least of Anadarko's challenges was getting to the drilling sites deep in the desert. A group led by then Algeria Operations Manager Carlos Mateus devised a plan to simply drive over the sand dunes — saving as much as $5 million in road-construction costs and putting in place a key piece of the puzzle that made it economically possible for Anadarko to justify staying in Algeria.

Making matters worse was growing political unrest, marked by violent demonstrations and the murders of foreigners in Algeria. Concerned for the safety of its employees, Anadarko closed the Algiers office in October 1993 and relocated staff to Houston. To more proactively communicate with the Algerian government, the company established a base office for the team in London, where the time zone was closer to Algeria's time zone. Under tight security, team members sporadically returned to the Algerian desert to continue their work.

These difficulties, in addition to the failure of the Berkine well, took their toll. Bob Daniels, a third-generation geologist from Wichita, Kansas, who had joined the Anadarko Algeria exploration team in 1992, recalls that "there were a lot of people in the organization who wanted us to sell our position, but John Seitz, who had taken over as the head of exploration in Algeria, was confident we would prevail in the long run."

Rather than quit, the explorationists went back to the drawing board. New 2-D seismic data revealed previously unseen fault trends, leading the team to conclude that their theories about the region's prospects were sound but their geological model was incorrect. They simply had drilled in the wrong spot. "The reservoirs were there," Chuck Abernathy, then Anadarko's vice president of operations and a member of the

Over Desert Sands

Vibroseis trucks send seismic signals through layers of earth beneath the sand dunes. A combination of intensive analysis of new seismic data, right, and instinct convinced Anadarko not to abandon Algeria after an initial dry hole.

Algerian exploration team, recalls. "We just had to figure out why the source and migration pathway had eluded us."

Daniels, Anadarko's current senior vice president of worldwide exploration, was instrumental in building a new geological model. "Bob mapped the areas that had the greatest potential for large discoveries," says Lee Petersen, tasked with pulling together a staff to figure out why the analysis on the Berkine well had been off the mark. "He put together an entirely new reservoir study, changing our focus from the Paleozoic to the Triassic zone and from the western part of the blocks to the eastern."

Petersen and his staff had three months to map the petroleum system. They pored over fault patterns, source rock, potential migration pathways and reams of other data in the interim. To obtain the most accurate subsurface images, they used state-of-the-art parallel computing and new Vibroseis seismic technology. Hopes rose for a big discovery. In February 1993, exactly three years and four months after Anadarko had signed its landmark contract with Sonatrach, the company drilled another well. The discovery wasn't big, but the team had found what it had come looking for.

Drilling at a depth of 12,300 feet at the El Merk block in the Ghadames Basin, the explorationists found hydrocarbons in two Triassic sandstone intervals. The first zone of the well flowed at a rate of 1,972 barrels of crude oil and 3.7 million cubic feet of natural gas per day. A second zone indicated similar pay.

In Houston, a standing-room-only crowd of employees assembled to hear Seitz read a press release announcing the find, a message that was simultaneously beamed via fax and satellite to wire services around the world. The throng erupted in applause and handshakes as Petersen and Mike Cochran, then Anadarko's manager of international exploration, poured champagne for a well-deserved congratulatory toast. "The oil flowed, the corks popped and everybody

Core Team

More than three years after signing its contract to explore Algeria, Anadarko celebrated its first discovery at the El Merk block in the Ghadames Basin. Opposite, from left, Bob Daniels, Richard Hook and Gary Ford, members of the Algeria exploration team, beam behind core samples from the Hassi Berkine North well, which led them to believe they would find huge fields in Algeria. Geologists study core samples, such as those at left, collected when drilling a well to learn more about the reservoir's characteristics.

cheered," *Currents* reported.

"You cannot imagine how excited we were about that first well," Allison says. "Although it was modest, it lifted our spirits tremendously."

Bonanza at Berkine

Revitalized, the explorationists kicked the drilling program into high gear. In February 1994, they announced two additional discoveries. The Hassi Berkine North well was the first significant discovery. It was the first well to encounter the TAGI formation — a Triassic sandstone interval full of oil. "The Hassi Berkine North well let all of us know that our models worked and that we were going to find some huge fields," reflects Daniels. The El Merk East well followed that success by encountering multiple zones and flowing more than 8,100 barrels of oil and condensate and 82 million cubic feet of gas per day. The Hassi Berkine North well flowed at roughly half these rates.

Every discovery offered information valuable to the next prospect. Still, the explorationists had yet to find the giant oil field they had pinned their hopes on almost five years earlier. They finally found it at the Hassi Berkine South field in March 1995 — a massive reservoir that at its peak produced 150,000 barrels of oil per day. When production came online in 1998, the field contributed 1.4 billion barrels to Anadarko's oil reserves. Within months of this momentous discovery, the team hit a second grand slam when they drilled the Berkine East well and discovered the Ourhoud field, which ultimately would produce 230,000 barrels of oil per day from an estimated 2.3 billion-barrel reservoir. Ourhoud represents Anadarko's biggest discovery in Algeria and one of the largest discoveries in the world in the last 20 years.

Into Production

When Anadarko discovered two giant fields in Algeria in 1995, plans were put in place to construct a central processing facility, above, in the desert to prepare produced hydrocarbons for market. Workers laid pipe, right, to transport crude oil to tankers for the trip across the Mediterranean to refineries in Europe.

The oil from Algeria was called Saharan Blend. It was clean and light and didn't require complex refining like thicker oils from Iraq or Iran. Aziz Abad, who had just joined Anadarko as its London-based shipping operations manager, recalls the sailing of the first ship carrying Saharan crude from Algeria. "We all flew to a port in Algeria called Skikda to celebrate as the ship, the *Orchid*, set sail on August 27," he says. "This was such fantastic news, to be involved in Anadarko's first waterborne foreign oil project. We had a great celebration."

Since that first discovery in Algeria, Anadarko's explorationists have found more than 2.8 billion gross barrels of oil and condensate producing 500,000 barrels of oil per day, roughly 40 percent of Algeria's daily oil production. This oil is refined to generate energy in myriad applications, providing incalculable benefit to humankind the world over.

Algeria was one of Anadarko's several international projects at the time. The company owned a 20 percent interest in a three million-acre lease block in Yemen and had signed an agreement to explore for hydrocarbons in 6.7 million acres in the Red Sea off the coast of Eritrea in Africa. Exploration also was underway in Peru, Indonesia and China, this last in 2.9 million acres of the South China Sea. Todd Fowler, a member of this exploration team, recalls his first meeting with geologists from the China National Offshore Oil Company, Anadarko's partner. "I'd been working at the Liberal office in Kansas, and when I pulled up to the hotel it was all very formal, shaking hands and bowing," he says. "The group then lined up and shouted, 'You're not in Kansas anymore, Toto!' I nearly fell over laughing." Richard Hook, another Anadarko geologist on the team, had put them up to it.

Desert Lighthouse

A lone derrick stands like a beacon in the harsh Saharan desert climes of Algeria. To date, Anadarko has discovered more than 2.8 billion gross barrels of oil and condensate in Algeria.

Hardly a "Dead Sea"

While Anadarko drilled many wells through its international ventures, the finds — with the exception of Algeria — proved modest to commercially disappointing. The company's offshore program in the Gulf of Mexico was a different story. By 1993, exploration drilling in the Gulf had dropped to a 20-year low, and some industry observers disparaged the body of water as a "dead sea." Geologists believed otherwise. They theorized that substantial oil- and gas-bearing structures were buried beneath the 140 million-year-old sedimentary salt formation extending 300 miles across the Gulf, from the outer continental shelf to the upper slope.

For decades, geologists had tried to solve the mystery of what lay beneath the salt, but they were hampered by an inability to understand the depositional models and salt tectonics. While conventional 2-D seismic data had produced a view of the top of the salt sheet, it yielded poor definition at the base. As had the Saharan sand, the salt in the Gulf distorted the seismic sound waves, producing the equivalent of snow on a TV screen — fuzzy data difficult to interpret. "Salt has twice the velocity of rock, so it distorts the seismic returns — making it very difficult to image the geology below the salt," explains

Under the Salt

Anadarko's technological prowess guided many discoveries in the sedimentary salt formation extending 300 miles across the Gulf of Mexico. Three of the company's most notable subsalt discoveries in the Gulf were Mahogany, Tanzanite and Hickory, all off the Louisiana coast. Anadarko's Mike Golden, left, studies a 3-D seismic image of Tanzanite. Below right is the platform at Tanzanite.

Stuart Strife, Anadarko vice president, exploration – Gulf of Mexico.

Furthermore, there were nearly insurmountable drilling problems. Companies that had drilled found nary a trace of hydrocarbons, compelling most to pull back or give up completely. Virtually alone, Anadarko pressed forward in pursuit of the Holy Grail — a view beneath the surface of the salt dome. In the early 1990s came a breakthrough, in the form of advances in 3-D seismic imaging.

Perhaps no play better symbolizes Anadarko's technological edge than its work in the subsalt trend of the Gulf of Mexico. Always a leader in the use of 3-D technology, Anadarko latched onto new tools such as pre-stack and post-stack depth migration analyses, which began to shed light on what lay under the salt. For the first time, explorationists could "see" beneath the large, horizontal salt sheets, and what they saw

Name Game

Explorationists take pride in naming their oil and gas discoveries. In the subsalt Gulf of Mexico, Anadarko's finds were named for different wood types, such as Mahogany, Hickory and Mesquite, above. Opposite, detail of a Mesquite wellhead.

were all the necessary components of an active hydrocarbon system.

In September 1993, the company and its partners, Phillips Petroleum Company and Amoco, announced the first commercial subsalt discovery in the Gulf of Mexico, tagged Mahogany. The well was drilled to a total depth of 16,500 feet in 370 feet of water, achieving initial flow rates of up to 3,700 barrels of oil and about 600,000 cubic feet of natural gas per day. Through additional perforations in the same zone, production ultimately reached 19,900 barrels of oil per day and 33 trillion cubic feet of gas. Anadarko held a 37.5 percent interest in the blocks.

The extraordinary find 80 miles off the Louisiana shore launched a new round of drilling in the Gulf of Mexico. At a federal offshore lease sale in March 1994, Anadarko invested $72 million in Gulf leases, all but two related to subsalt. Within months, the company had spud two additional subsalt wells — the Teak and Mesquite prospects, in which Anadarko owned a 50 percent working interest. Additional exploration divined other reservoirs over the years, such as Tanzanite, Hickory and Agate. It was a remarkable series of finds that earned a profile of Anadarko in *Forbes* magazine. The headline said it all: "The Gulf of Mexico is No Longer a Dead Sea."

Anadarko's high-risk, high-potential triumphs in Algeria and the Gulf of Mexico balanced its more stable developments and commercial disappointments. "Everything we do requires a proper risk profile," Allison says. "You have to have bread-and-butter, low-risk plays and a certain amount of medium stuff. But you also have to take high-risk, high-potential bets in areas like Algeria, which turned out to be hugely successful, and Eritrea, which was unsuccessful." His strategy in high-risk plays was to eschew a conservative approach. "The key is to not tread lightly with a one-well deal," he explains. "You want a deal where you learn something from the first well, which improves your odds on the second, third and fourth wells. In this business, you have to give yourself a chance to be lucky."

Tenacity Pays in Alaska

Anadarko extended its run of good luck in Alaska, one of the last domestic frontiers for oil and gas exploration. With partner ARCO Alaska, the company explored the state's North Slope for hydrocarbons. In 1996, the partners made their first big find at the Alpine field, about 50 miles west of the Prudhoe field, the largest field in the state. The Alpine reservoir contained proved oil reserves of 430 million barrels — the industry's

biggest domestic onshore discovery in the last 10 years.

The discovery almost never happened. "Wells had been drilled in the Alpine field area in the past by Texaco and others. Oil was found but not in commercial quantities," says Kevin Martindale, project geophysical advisor. "ARCO wanted to stop drilling and plug the well after the shallower objective was dry." As it had done in the Gulf of Mexico and Algeria, Anadarko held tight to a prospect that other explorationists shunned. "John Seitz insisted that we drill deeper to the original target depth we'd agreed upon, which caused some commotion," Martindale notes. "The partnership ended up agreeing to drill the entire section. And that's when we made our first important find with the Bergschrund well." Production from Alpine peaked at more than 130,000 barrels per day in 2005, and the field is still producing at near-peak levels.

Alpine isn't just a story of persistence and luck. The field is considered a showcase for

North Slope Alaska

Just a few years out of college as an Alaska exploration geologist, Jennifer Burton was on site at a North Slope exploratory well to interpret well information. Anadarko and partners had to build ice roads to haul rigs and other equipment. They used specialized vehicles called rollagons, left, to move across the tundra without causing harm. At the snow-covered Alpine field on Alaska's North Slope, opposite, horizontal wells were drilled from two specially constructed surface pads covering a small portion of the total field area, reducing the impact on the ecosystem.

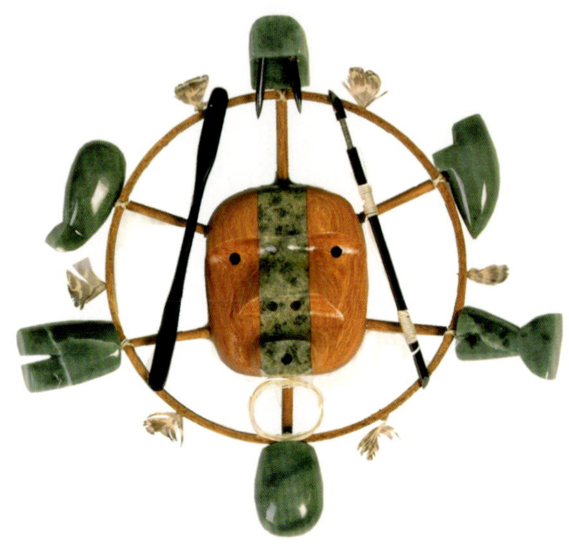

Peak Production

Production at the Alpine field, above, peaked at 130,000 barrels of oil per day and continues at peak levels. State-of-the-art technology, right, is used to manage the field's input. The spirit mask, above right, was carved by a native of Elim, Alaska, and presented by the Alaska Support Industry Alliance to Senior Vice President Mark Pease. Opposite, an exploratory rig on Alaska's North Slope.

environmental sensitivity. To minimize the impact on wildlife and the surrounding ecosystem, ARCO and Anadarko drilled long-reach horizontal wells from two surface pads that covered 100 acres — less than one-quarter of 1 percent of the total field area. The companies built a small gravel airstrip to fly staff to and from the field, eliminating the need for roads and bridges. To prevent disruption to the Colville River, they buried the pipeline used to transport the oil 100 feet below the riverbed. In many ways, the field was developed like an offshore platform — a standalone facility without roads. "We purposely built a production facility with a very small footprint at the Alpine field, doing a lot of innovative engineering to figure out how our operations could coexist with wildlife in such an environmentally sensitive area," says Martindale.

Alaska is an important frontier play for Anadarko. The company signed a strategic alliance with ARCO in 1996 to explore the Upper Cook Inlet on the state's southern coast. The partners later acquired six blocks totaling 34,000 acres in the Beaufort Sea. Anadarko also signed an exclusive agreement in 1998 with the state's

Crowded Lot

Frac trucks, above, fitted with diesel engines and high-pressure pumps, inject fluids at a high rate into the reservoir to release natural gas trapped in the Haley field. Neil Labbe, right, production superintendent, helped bring the County Line coalbed methane project in Wyoming online in the 1990s. Opposite: silhouette of a worker at the Atlantic Rim coalbed methane field in southern Wyoming.

largest native corporation to explore more than 2.2 million acres in the foothills of the North Slope, and in 1999 expanded its acreage position by acquiring 99 tracts in a federal lease sale of properties in Alaska's National Petroleum Reserve. That same year, ARCO and Anadarko announced an important discovery at the Fiord field, a satellite to Alpine boasting potential gross reserves of more than 50 million barrels of oil.

Next Frontier: Coalbed Methane

Such frontier discoveries were in addition to the company's unconventional plays, including the first coalbed methane project at the Helper field in Utah. Anadarko had proven its skill extracting hydrocarbons from other unconventional reservoirs with tight gas sands or shale formations. It saw coalbed methane as a natural extension of this work. Unlike conventional reservoirs in which gas is contained in the porous spaces within the rock and is easily released, coalbed methane reservoirs contain gas that literally clings to the coal in a powerful molecular attraction. The gas has to be freed from the coal so it can rise to the surface.

In spite of the technical and economic challenges, Anadarko was determined to find a way to separate the gas from the coal, largely because of the potential payoff — a coalbed methane reservoir can contain two to three times as much gas as a conventional one. Anadarko cracked the code after an exhaustive analysis of the physical properties of the coalbed. "There is a natural fracture network in coal called 'cleats' that contain water," explains Jerry Windlinger, vice president of U.S. business development and a member of Anadarko's coalbed methane group in the 1990s. "When you pump this water out of the cleats, the pressure drops, causing the natural gas to bubble up to the surface over a period of time."

The downside of this method is that it can take two years or more for the gas to rise, slowing recovery times along with the rate of return. Countering this drawback, however, are very low

finding and development costs. Windlinger notes that "from a portfolio perspective, coalbed methane prospects balanced our other prospects that had a high cost of finding and development and a great rate of return."

Helper field started producing in 1993 and reached 72 million cubic feet of gas per day at its peak. The success of the project guided another major coalbed methane play in Wyoming's Powder River Basin, which currently produces about 145 million cubic feet of gas per day. Similar prospects were developed at the Drunkards Wash field in Utah and the Big George field in Wyoming. Coalbed methane is an important Anadarko asset; holdings have increased from 50,000 acres in 1999 to more than one million acres at the end of 2005. "In the mid-1990s, we were picking up 10-year leases for an average of $13 an acre in Wyoming and Utah," Windlinger says. "It was a good deal."

Bossier's Tight Sands

Such unconventional plays became an Anadarko trademark. The company was determined to squeeze oil and gas from prospects long overlooked or abandoned, using emerging technologies and innovative processes. Knowledge and expertise gleaned from one unconventional play were applied to the next. Anadarko's experience in central Oklahoma's Golden Trend, for instance, gave it the drilling and completion know-how to later attack the tight sands of the Bossier Trend in Freestone County, Texas.

Bossier presented a unique challenge: fracturing the rock and drilling wells were prohibitively expensive. Bossier was known to hold great reserves, but as Steve Pearson explains, "The rate of return was limited. You had to hire a fracturing company that used a gel-based substance mixed

OBI

All Together Now

When Anadarko realized that anticipated production from Algeria would not arrive in time to meet the company's oil production target in 1997, employees came to the company's rescue, devising ingenious ways to cost-efficiently boost production at domestic properties. Employees called their ambitious project "Sudden Impact," the name of a popular movie starring Clint Eastwood.

Rex Alman, former Anadarko senior vice president of onshore operations, became the project's unofficial ambassador. "I traveled to all production offices and put my phone number on the wall and said, 'Any time you young engineers have an idea to build production, and management is giving you static, call me directly and we'll see about getting the project approved,'" Alman says.

To ensure that employees knew that Alman was serious, Anadarko's Steve Martin created a stamp for Alman to approve new projects, emblazoned with the letters OBI, for "Obvious by Inspection." "It meant that if someone had a project and it was obvious by inspection that this project had merit, we wouldn't waste a bunch of time running fancy economic models before we gave it the thumbs-up," Alman explains. "Time was of the essence."

In the meantime, Bob Allison added pressure of his own. "I said, 'Why don't we sell the West Texas operations — we're not getting anything out of them.' When people heard that, they came up with lots of prospects. I never intended to sell. I just wanted them to perform."

To attain the goal of 44 million energy equivalent barrels, employees challenged themselves to squeeze everything possible out of Anadarko's fields. The entire company was in fast forward. An average of 28 company-operated rigs ran at a near-constant hum, spudding 756 wells in 1997 compared to just 13 rigs spudding 358 wells the prior year. Employees celebrated when they hit the goal. Alman ordered T-shirts, mugs, thermoses, hats and other merchandise sporting a logo emblazoned with the number 44. "We all pulled in the same direction at once and it made a huge difference," he beams, a baseball cap with the number "44" perched squarely on his head.

Collaboration toward a common goal — reaching Anadarko's stated oil-production target of 44 million barrels — marked the successful Sudden Impact project in 1997. Employees joined hands to devise innovative strategies to boost domestic production; their success was commemorated with a variety of merchandise emblazoned with the number 44.

with some sand that was pumped into the rock to fracture it wide enough to get the gas out. This gel cost quite a bit of money. No sooner had we gone into Bossier than we backed out, figuring we couldn't make it commercial."

Anadarko's partner in exploring the Bossier field at the time was Apache Corporation. When the cost of drilling wells using the gel substance exceeded the field's commercial potential, the company farmed out its interests to Apache. It was a premature decision. Pearson and a small group of Anadarko engineers, geoscientists and petrophysicists continued to study ways to improve the economics of drilling in the tight gas sands — and they came up with a simple idea: eliminate the gel-sand mixture and just pump water into the rock. "We figured the tight sands would still crack and let the gas escape," Pearson says. "Just because we were dealing with sand didn't necessarily mean the cracks would close. If you break a brick in half and put the brick back together, you can still see the crack."

The team based its assumptions on a process called hysteresis, in which the original properties of an object, once undone by some external influence, cannot be restored. Robert Talley says Pearson's work showed that the company could substantially lower its development costs, and that finding led Anadarko back into the project.

When, in 1996, Pearson and his team finally fractured Bossier's tight sands with nothing but H_2O, the gas surprisingly flowed even better than it had following conventional gel and sand fracturing. The price tag for the fracturing service was about $50,000, about one-third the cost of the gel-sand process. This, along with the use of more durable and efficient PDC (polycrystalline diamond compact) drill bits, helped reduce drilling and development costs from $3 million per well to $1 million per well. "We completed the wells using just water fracs and suddenly we had a booming play," Bob Allison recalls.

Anadarko increased its gross acreage position in the Bossier Trend to more than 250,000 acres in succeeding years and reached a 2004 production milestone of 300 million cubic feet of natural gas per day. The company also acquired leases on more than 600,000 acres in northern Louisiana's dormant Vernon field, another tight gas play forsaken by the industry. "We took the ideas and techniques we learned at Bossier and looked for other places where we could successfully leverage our expertise," Mark Pease says. The Vernon field reached a production peak of 350 million cubic feet of natural gas per day in 2004. Another field left for dead had come alive.

Sudden Impact

When Anadarko celebrated its 10th anniversary as an independent public company in 1996, there was much to cheer. The company had doubled reserves over the last decade to more than 600 million energy equivalent barrels, and for 15 years in a row had replaced production with new reserves — the only oil and gas company in the industry to achieve this distinction. Its financial condition was equally robust. Net income had grown tenfold, from $10.1 million in 1986 to $100.7 million in 1996. Those first 100 shares that Bob Allison bought for $22 a share now traded in the $65 range.

As good as 1996 was, the next year was even better, marked by new discoveries, increased reserves and a cost-of-finding rate that, at $3.15 a barrel, was the lowest among independent exploration and production companies. In the meantime, Anadarko had evolved from a domestic natural gas company into an international oil and

Leader with Impact

Rex Alman III led Anadarko's Sudden Impact project. Alman began his career at Anadarko in 1976 as an evaluation engineer in Houston and became senior vice president of domestic operations and then Algeria before he retired in 2003.

gas enterprise. Crude oil now represented more than half the company's reserves, and Anadarko boasted ventures in major overseas basins across the globe.

Allison was resolute about sustaining international exploration and development. He pledged $176 million to this effort in 1996, marking a 227 percent increase from the previous year's budget. The company explored the North Atlantic Margin, acquiring leases on 1.3 million acres offshore Ireland and Scotland. It also hunted for hydrocarbons in Tunisia on 384,000 acres near the Algerian border in the oil-rich Ghadames Basin. Anadarko's explorationists hoped to leverage the knowledge and experience they had gleaned in Algeria to the Tunisian sector. "Technically, we understand this area as well as anybody in the world," says Bob Daniels.

Algeria remained Anadarko's most bountiful overseas play, but the company confronted the discomforting possibility in 1997 that 1.5 million barrels of production from these properties might be delayed. Without the production from Algeria, Anadarko was in danger of missing the growth forecast it had pledged to investors for that year — 44 million barrels of oil equivalent — and the company wanted to maintain investors' confidence by showing continued double-digit production growth.

Safety First

Anadarko workers help their colleagues gear up for hazardous materials emergency response training — part of the core safety training program at Anadarko, where employee safety is a priority. Opposite, a worker makes the daily rounds of production assets in the Permian Basin of West Texas, long an Anadarko stronghold.

Determined to drive up production, Anadarko employees pulled together in a program called "Sudden Impact" to hit the target Allison had pledged to Wall Street. Most of the work fell on Anadarko's U.S. onshore region, the only operations that could realistically add production quickly enough to have the required effect. Rex Alman was head of U.S. onshore operations at the time. He says employees kicked themselves into high gear, devising hundreds of projects that, in many cases, had little impact on their own but when added together had a large and immediate impact.

Production workers began laying pipe while wells were still being drilled instead of waiting for them to be tested, accelerating production from new wells by months. They installed systems to collect small amounts of natural gas that previously had been vented from the tops of oil tanks. Compressors were used to suck gas out of low-volume wells. Reservoir and geological workers reduced the time it took to identify drilling locations; landmen accelerated their permitting processes and accountants worked with them to book production to wells faster. Employees posted record levels of activity in the Hugoton Embayment, the Permian Basin and East Texas.

"Everyone in the company understood the goal and worked very hard to ensure that we achieved it," Alman says, clearly proud that Anadarko hit the target. "That is why you still see '44' on so many things around the company."

"Dance with the One that Brung Ya"

While the mood was upbeat at Anadarko, the oil and gas market slumped further in 1998. Natural gas prices were down almost 20 percent from 1997 levels while crude oil prices tumbled nearly 40 percent over the same period. In real dollar terms, oil prices were the lowest they'd been since the Great Depression. Three factors conspired to wreak havoc — a growing economic crisis in Asia, a warmer than normal U.S. winter and increased oil production, much of it from the Organization of Petroleum Exporting Countries (OPEC). There simply was too much oil on the market.

Most companies in the industry reacted by slashing budgets and employment. Hundreds of oilfield development projects were delayed, deferred or canceled, and more than 65,000 workers in the oil patch lost their jobs. Not at Anadarko. "Bob always said, 'You dance with the one that brung ya,'" recalls Julie Struble, Anadarko vice president, operations – southeastern region. "He felt that if people bring you success, you keep dancing with them — you don't let them go."

Allison earmarked capital to continue developing Anadarko's giant fields in Algeria, Alaska and the Gulf of Mexico, though the level of spending was down slightly from prior years. As always, he was pragmatic. "Like most commodities, oil and gas prices move in cycles," he says. "Our people are adept at quickly capturing new opportunities in our core areas while maintaining a competitive ability to move into new plays. We've proven even when times are tough we can grow production and reserves."

Despite these strengths, Anadarko and the rest of the exploration and production sector were beset by sharply lower financial returns in 1998. Anadarko's revenues fell from $675 million to $560 million, and the company posted a net loss of $42 million. Cash flow was severely impinged, challenging efforts to aggressively pursue a broader portfolio of projects — an escalation of the company's historic strategy of growing through the drill bit. For a company that had grown like quicksilver since its inception, the financial setback required new thinking.

Anadarko began to seriously consider acquiring another major E&P company. Looking around for the right one, the company spied a prized catch.

CHAPTER FIVE: BIGGER VISION THAN WE DARED ON OUR OWN

Career Couple

Ellen and Gary Cobb both enjoyed long careers at Anadarko, beginning with the company's early days as a subsidiary of Panhandle Eastern in Liberal, Kansas, Gary's hometown. Gary retired as a computer systems analyst in 1999, Ellen in 2003 as a senior marketing analyst. "We car pooled to Anadarko for 33 years," says Ellen. "At work, we never ate lunch together. We stayed totally apart, which was a good thing for our marriage."

The jewel that Anadarko had spotted was Union Pacific Resources, formerly a subsidiary of the giant Union Pacific Railroad. In July 2000, when Anadarko acquired UPR for $4.3 billion in stock, the size of the company doubled overnight.

The acquisition of UPR made Anadarko the largest independent E&P company in North America in terms of market capitalization, worldwide cash flow and reserves. Although the companies held comparable reserves, Anadarko had reeled in a partner whose revenue stream was twice as large. This ironic story of a smaller fish eating a bigger one earned the merger headlines in dozens of major business publications worldwide.

Post-merger Anadarko's market capitalization exceeded $9 billion, its global land holdings encompassed more than 21 million net acres and its daily energy output comprised 1.7 billion cubic feet of natural gas and 203,000 barrels of oil. Anadarko was now the leading acreage holder, the most active driller and the company with the most production of any of the independents in North America. Its U.S. gas production and reserves were so plentiful that they surpassed those of better-known rivals like Texaco, Phillips and ARCO. To veteran employees like computer systems analyst Ellen Cobb, who'd been with Anadarko since the beginning, this new-found range and renown bordered on the miraculous. "I was just so darn proud," says Cobb.

Unlike many mergers in the oil patch, this one wasn't predicated on cost savings. Rather, the merger was complementary. UPR had been struggling with debt, the legacy of a failed hostile attempt to acquire Pennzoil in 1997 and the successful but costly $3.6 billion takeover of Canada's Norcen Energy Resources in 1998. UPR paid a premium for Norcen just as energy prices collapsed. It also assumed $900 million of Norcen's outstanding debt, racking up more than $2.7 billion in debt. "We were so busy paying down debt, we left many prospects unevaluated, unexplored and undeveloped," says David Larson, Anadarko's former vice president of investor relations.

Despite these distractions, UPR was generating more than $1 billion annually in cash from its various oil and gas assets. Anadarko, meanwhile, was crimped for cash by the slump in oil and gas prices. In acquiring UPR, it would obtain access to capital to aggressively expand worldwide exploration and development, achieving greater growth and profitability than it could on its own. But cash flow was just one inducement. UPR's drilling and completion technology was unparalleled. It was particularly adept at horizontal drilling and had put this expertise to profitable use in East Texas' gas-rich Austin Chalk field. UPR also was unquestionably the most active driller in the United States, running more onshore rigs than any other E&P company. One trade publication labeled it a veritable "drilling machine."

By blending UPR's drilling and completion expertise with Anadarko's industry-leading exploration capabilities, two successful companies became a single, better company. "It was an excellent fit, greatly improving what either one could have done alone," says George Lindahl III, CEO of UPR at the time of the acquisition. "The

Chalk it Up

Union Pacific Resources was the undisputed master of horizontal drilling in the 1990s. This technique, in which a well drill bore is turned sideways to open up more of a reservoir for production — was just what the doctor ordered at central Texas' gas-rich Austin Chalk trend.

The play had experienced the boom-bust characteristics typical of many mature reservoirs and had lost commercial appeal for most of the oil and gas industry. "It was known as a 'promoter's paradise,' meaning you rolled the dice on it," says Jeff Stahley, an engineer at UPR at the time and today an Anadarko onshore operations manager. "Sometimes you drilled a well in the chalk and, if you hit a fracture, you got a great well. Other times you drilled and you didn't hit a fracture and got a real dog of a well. You'd always get a little something, but some wells were good and others never paid out."

Like the rest of the industry, UPR was fed up with the Austin Chalk, so named because of its limestone and marl formations. "We tried to get out of it and put the whole thing up for sale, but nobody wanted it," Stahley says. Stuck with the asset, UPR tried a different tack. "We had a couple of guys playing around with horizontal drilling who figured they might be able to intersect more of the fractures by applying the technology to the chalk," Stahley says. "They learned that the fractures lined up in patterns with a northeast-southwest trend to them. If you drilled a horizontal well perpendicular to these fractures you had the best chance of intersecting them."

UPR drilled its first horizontal well in the Austin Chalk Gittings field in 1989. "That first well turned out to be a good science project," Stahley says. "It wasn't a great success, but what we learned from that well helped us achieve great success in the play."

Within a year, the company had 30 drilling rigs humming at the site. "Pretty soon we had landmen leasing land all across the chalk," says Stahley. "It grew to where we have around 1,500 active wells there today. It's an example of how technology and skill can successfully be applied to dubious prospects that seem to have no life left in them."

Anadarko's production today from the Austin Chalk trend is in the range of 50,000 barrels of oil equivalent per day.

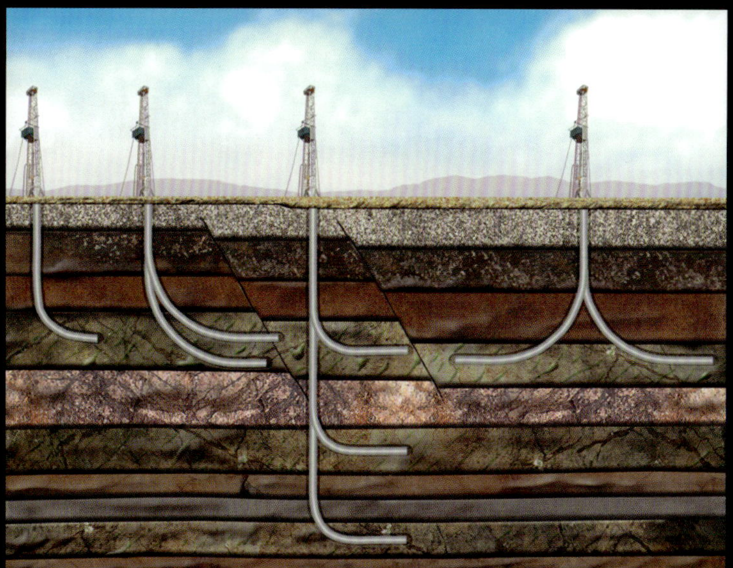

Of the many assets the acquisition of Union Pacific Resources brought Anadarko, the Austin Chalk trend was a particular jewel. UPR earned industry recognition by adeptly applying its horizontal drilling expertise to the field, which boasts 1,500 wells today.

merger was all about complementary skills, portfolios and assets, creating one company with a vision of growth and opportunity — a vision bigger than either one of us dared imagine on our own."

UPR's Land-Grant Bonanza

Also persuading Anadarko to buy UPR was the latter's huge land position, which included prime mineral and hydrocarbon reserves in Colorado, Wyoming, Texas, Louisiana, Canada and South America. The merger brought Anadarko a more balanced portfolio both geographically and geologically, diversifying its producing properties and exploration prospects.

UPR's origins derive from the development of energy reserves on land granted its former parent, Union Pacific Railroad, under the first Pacific Railroad Act, which was signed by President Abraham Lincoln in 1862 (and modified many times until 1874). The land grant was the government's way of loaning capital to the railroads to provide transportation and telegraph links between the Missouri River and the frontier

Rocky Potential

Anadarko owns mineral rights to millions of acres in the Western states of Wyoming, Colorado and Utah. The Western states are rich with natural gas and have become a major focus of Anadarko's domestic exploration efforts. The company drilled almost 300 wells in 2005, with a 100 percent success rate. The Helper and Drunkard's Wash fields in eastern Utah are important sources of coalbed methane, producing almost 57 million cubic feet of natural gas per day in 2005. The sun highlights Drunkard's Wash gas compression facility against the backdrop of the Rockies, left.

Lincoln's Largesse

The saga of Union Pacific Resources is keenly interwoven with the settling of the Wild West. The story begins nearly 150 years ago with the grant of a huge swath of land to Union Pacific Railroad. The U.S. Congress and President Abraham Lincoln awarded the land to Union Pacific and other railroads following passage of the Pacific Railroad Act of 1862. The law and its subsequent revisions provided incentive to railroad owners and investors to build the Missouri River-to-California segments of the transcontinental railroad, along with a coast-to-coast telegraph system.

Lincoln and his supporters believed a web of railroads to be vital to the national defense and ongoing development of the western territories. They saw the Railroad Act, passed during the height of the Civil War, as a way to perhaps curry loyalty from California and Utah to the Union. The government issued millions of dollars in bonds to construct the railroads and essentially loaned this cash to them. The land grant was in effect the collateral on these bonds. It was a bold financial concept: once commerce was established in the towns along a rail line, the railroads would be in a position to sell the land and reimburse the government.

Union Pacific received a grant of half the land on a strip 20 miles wide on both sides of the track for each mile of railroad it constructed, approximately 6,400 acres of land per mile of track, or 11.3 million acres in total. The other half of this land was owned by the federal government, resulting in a

Union Pacific Resources was built upon the mineral deposits owned by its parent company, the Union Pacific Railroad. Coal initially was the predominant resource, particularly the high-quality coal mined in Rock Springs, Wyoming, opposite. Miners such as the one above left and hard at work underground at Rock Springs, have always been a hardy lot.

checkerboard pattern of land ownership along the track that ultimately stretched from Omaha, Nebraska, through Kansas, Colorado, Wyoming, Oregon and Montana to a junction in Ogden, Utah, with another line being built by the Central Pacific Railroad.

"What drove the route of the railroad were Indians, water, coal and lumber," says George Peters, an Anadarko landman whose career began at UPR in 1977. "The railroad wanted to avoid trouble with the Indians, but they also needed access to water for steam and to coal and lumber for fuel. The reason many towns along the railroad are spaced 13 miles apart is because that's how far a locomotive engine could travel before it needed water."

While farmers and cattlemen bought most of the land, a sizable share contained significant coal deposits, particularly in Rock Springs, Wyoming, where the high-quality coal was considered a longer-lasting source of energy than coal back East. Union Pacific negotiated agreements with several coal-mining companies to develop the reserves before deciding to exploit the coal itself in the mid-1870s, via a subsidiary, Union Pacific Coal Company. Today this same coal produces electrical power for millions of homes and businesses in the western United States.

Anadarko holds approximately eight million acres of UPR's former mineral rights in Colorado, Wyoming and Utah. The company's minerals group promotes the development of these resources, which provide more than $42 million a year in royalties, helping fund the growth of the company's oil and gas operations. It is an extraordinary expanse of land, rich in resources of trona, limestone, titanium, zeolite, oil shale, diamonds — and, of course, the coal that once powered fuel-hungry locomotives transporting people and resources to the still-wild West.

West. Union Pacific sold much of its land to agricultural, mining and ranching enterprises to raise cash to repay the government. The land remaining in the railroad's control contained significant mineral and hydrocarbon deposits.

While Union Pacific was aware of large deposits of oil and natural gas in its Utah and Wyoming land holdings, its initial attempt in 1872 to develop these reserves in Wyoming's Green River Basin as a locomotive-engine lubricant proved unproductive. Another 50 years elapsed before the railroad revisited commercial development. It was the emergence of the automobile industry, which fostered demand for refined products like gasoline, that revitalized Union Pacific's interest. In 1937, the railroad finally gauged the size of its hydrocarbon reserves and drilled 72 wells, all of them producers.

To manage a now-burgeoning enterprise, the company formed an oil and gas subsidiary, Union Pacific Land Company. Upland, as it was called, averaged annual net earnings of $30 million between 1946 and 1960, and by the 1990s produced more than one million barrels of oil per month. The federal subsidy to build the Union Pacific Railroad had emerged as a bonanza of exceptional proportions.

Upland had one problem, however. It lacked the exploration and drilling expertise to fully exploit the railroad's oil and gas resources. In 1969, Union Pacific decided to farm out exploration and drilling on sections of its land holdings to Amoco, receiving royalties in return.

Sharing proceeds with Amoco was not the most financially rewarding strategy. A better idea emerged for the long term — the purchase of an oil and gas company with broad skills filling Upland's void. In 1970, Union Pacific acquired Champlin Petroleum, an integrated oil and gas company with a range of assets: international exploration, production, refining, processing and the sale of refined oil products through 1,300 service stations from the Great Plains to the nation's heartland. The acquisition further served as a natural hedge to the price of transportation fuel, making it a winner on many levels.

Champlin had been founded in Enid, Oklahoma, in 1916 by H. H. Champlin, a banker who drilled a well on land leased for $11,500 and discovered a reservoir gushing 300 barrels of oil per day. Champlin purchased a small refinery in 1917 to develop these reserves and thereafter built pipelines connecting Enid with Oklahoma City and neighboring Midwest states. In 1923, he opened the first Champlin service station in Enid.

Union Pacific folded Champlin into its minerals subsidiary, Rocky Mountain Energy, in 1987 and created a new name for the combined entity — Union Pacific Resources. By then, the Champlin gas stations and refineries had been sold, and UPR concentrated strictly on exploration and production. The company was spun off to shareholders in 1995 as part of Union Pacific's strategic decision to focus exclusively on railroads.

Three years hence, UPR acquired Norcen to gain a stronger foothold in Canada, a country whose oil and gas reserves exceeded those of the lower 48 states. Norcen also was a company with deep roots, having evolved from a series of mergers undertaken by Pathfinder Petroleum, an oil and gas concern incorporated in 1952. Norcen was active in the exploration, production and transmission of oil and gas in Canada, and it held a significant position in the Gulf of Mexico's continental shelf.

Timing is everything in acquisitions, and with UPR's purchase of Norcen, it couldn't have been worse. Oil prices collapsed, falling 25 percent below the prices in effect at the time of the acquisition. "UPR's balance sheet was under pressure after the Norcen deal because of lower cash flows, and we had to sell our gas processing and marketing businesses to pay down the debt," says Larson. "The acquisition was strategically sound, but it ultimately wounded us. We over-leveraged ourselves, thinking we could pay down the debt. Then oil fell to $10 a barrel."

Fortunately, different market dynamics greeted the merger of Anadarko and UPR: oil and gas prices were heading up after more than two years of sagging. The *Wall Street Journal* commented that the combined companies "will have the muscle to play a major role in the North

Swaths of Land

The land granted to the railroads by President Abraham Lincoln was predicated on the building of a transcontinental railroad linking East and West. The grant gave the railroads ownership of a wide swath of land — and thus valuable mineral rights — in a checkerboard pattern on either side of the proposed rail bed.

Whipping up a Deal

The merger of Anadarko and Union Pacific Resources in 2000 was an on-again, off-again affair challenged by human frailty and Mother Nature.

In 1998, Bob Allison and Drew Lewis, the CEO of UPR at the time and a former U.S. Secretary of Transportation in the Reagan Administration, met at New York City's venerable Waldorf Astoria Hotel to consider the merits of combining the companies' complementary assets. The negotiations were put on hold, however, when Lewis' health deteriorated.

Allison revived the discussions at an industry gathering in Scottsdale, Arizona, in February 2000, in a meeting with George Lindahl III, Lewis' successor. "Bob sent his investment banker to see me and asked if I was open to talking," Lindahl recalls. "I agreed to meet him, but I had other plans for the company at the time."

Lindahl had hoped to resuscitate UPR independent of a merger, but he soon realized that a deal with Anadarko was in the best interests of shareholders. "My one goal was to create shareholder value," he explains. "Other companies courted us, including several independents, but Anadarko had the best stock multiple of them all. Bob and I met and he offered to pay a premium for UPR at a time when oil and gas prices were turning up. The deal was too good to pass up."

The two CEOs were a study in contrast. Allison is the picture of executive restraint, a conservative man comfortable in tailored suits and casual attire at business conferences. Lindahl sports a well-trimmed beard, leather pants and close-cropped hair, and favors Hawaiian shirts, Panama hats and a deep tan. Yet both men found mutual ground to make a deal and, after Lindahl became Anadarko's vice chairman post-merger, they worked together closely and smoothly.

Then-CEOs George Lindahl, left, of Union Pacific Resources, and Bob Allison of Anadarko brokered the merger of their companies in 2000. The transaction doubled Anadarko's proved reserves.

The deal almost didn't come off a second time. A tornado shut down UPR's Fort Worth offices on the final day of negotiations; headquarters was heavily damaged by the twister and was closed for more than two weeks. Executives had to hammer out last-minute details at a hangar at Fort Worth's Meacham Airport and the Mira Vista Golf Club. "I remember we had to go back into the offices in hard hats after the tornado to get some documents that were important in working out some issues," Lindahl recalls. "We then moved to the hangar, which had a conference room and some offices with Internet access to our network. I think we squeezed like two dozen people into a room that no more than eight or nine people should have been in."

The deal was approved by UPR's board at the Mira Vista two days later. Lindahl later left Anadarko and is now managing partner of Sandefer Capital Partners, a Houston-based oil and gas investment company. He still has his tan.

American natural gas market at a time when prices are holding at healthy levels."

UPR's future is now entwined with that of Anadarko's. The land grant on which the railroad system and the American West were built is part of Anadarko's legacy. "Few companies have such an important historical origin," says George Peters, an Anadarko landman. Moreover, the eight million acres of land holdings remaining represent some of Anadarko's most prized mineral rights.

Anadarko's acquisition of UPR was not its last. In the ensuing years, the company would engineer several more acquisitions, giving it extraordinary range, size and intellectual capital. The cost of these mergers would prove formidable, however, making Anadarko itself the subject of takeover rumors.

Canadian Production

Dave Kapple joined Anadarko in a summer job in 1979. A long-time Anadarko engineer who recently oversaw assets in Canada, he is currently production engineering manager – MidContinent/West Texas. Left, an Anadarko drilling rig in British Columbia, Canada.

CHAPTER SIX: SOARING — THEN BACK TO EARTH

Anadarko rang in the new millennium on a high. The merger with Union Pacific Resources produced tremendous shareholder benefit, exceeding the value achieved by any other acquiring company in any industry in 2000. The market value of Anadarko's shares more than doubled, while the shares of its peer group in the oil and gas industry increased by only a third, on average. Many analysts extolled the deal as the most successful corporate merger of the year.

Anadarko also benefited from the energy market's dramatic rebound. Oil prices nearly tripled and natural gas prices soared to record highs in 2000. Anadarko's reserve replacement of 213 percent drove proved reserves to the highest levels in its history — providing ample room to capitalize on the market's revival. It was the 18th year in a row in which reserve adds exceeded production. By 2001, annual production skyrocketed to 199 million barrels of oil equivalent. Buoyant business boosted earnings and cash flow to record highs, giving Anadarko the financial capital to grow both at the drill bit and through acquisition.

Lance Lauck, head of Anadarko's merger and acquisitions group, and his team examined several merger candidates, among them Berkley Petroleum Corporation, a Canadian exploration and production company. Berkley's portfolio included 1.5 million undeveloped acres for exploration and operations spanning Western Canada's Sedimentary Basin in Alberta, British Columbia, the Northwest Territories and Saskatchewan — high-impact areas in Western Canada where Anadarko had no critical mass and sought a larger presence. "Berkley also would bring balance to our exploration risk profile, offsetting

Norcen's low-return, bread-and-butter projects with more medium risk-medium return plays," Allison notes.

Berkley held an estimated 100 million barrels of oil equivalent of net proved reserves, of which 70 percent were natural gas. The company was a major player in Alberta's Wild River Basin, which held an estimated four billion cubic feet of natural gas reserves from multiple pay zones in each one-square-mile section. Berkley had drilled 30 wells that produced approximately 27 million cubic feet of gas per day, and the widespread feeling was that it had only cracked the surface. With gas prices flying high, Anadarko saw an exceptional opportunity to acquire a well-positioned company on the financial upswing.

So did Anadarko's rivals. In December 2000, Dallas-based Hunt Oil Company made a hostile bid for Berkley. To fend it off, Berkley's board of directors sought bids from other suitors — and those bids would be due in six weeks. Anadarko quickly assembled a multi-disciplinary team of engineers, geologists, geophysicists and landmen to flesh out a competitive offer. For two intense 12-hour days in Berkley's Calgary headquarters, the team pored over the company's engineering and geological records, maps and financial statements. They made the deadline by a whisker. On March 16, 2001, Berkley accepted Anadarko's bid of approximately $1 billion in cash and the assumption of Berkley's debt.

Building Bridges

Anadarko took what it had learned from its hydraulic fracturing experience at the Bossier field and applied it successfully, left, to the Wild River Basin's tight gas formations in Alberta, Canada. A roughneck, opposite, works a rig. As Anadarko vice president, Canada, Mike Bridges, above, oversaw exploration and development of the company's holdings across Canada. These days, he serves as vice president, operations – Northern Rockies.

A Wild Ride

No sooner had the companies merged than Anadarko plunged head-first into the Wild River, drilling 70 wells that hiked production to 90 million cubic feet of gas per day by the end of 2001. The company then applied its hydraulic fracturing expertise from the Bossier field to Wild River's tight gas formations. Production rates jumped 83 percent and well completion efficiency rose 50 percent. Total Wild River production eventually hovered around 105 million cubic feet per day from 250 producing wells. "This was our most important asset in Canada," says Mike Bridges, then-president of Anadarko Canada.

The ink was barely dry on the Berkley contract when Anadarko acquired another Canadian company — Gulfstream Resources, an international oil and gas exploration and production company based in Calgary. Anadarko funded a portion of the $137 million purchase price by selling properties in Guatemala and Argentina, part of the company's strategy to swap lower-margin, non-core assets with prospects offering higher value and greater growth. The deal marked Anadarko's entry into production and exploration in the Middle East, a world-class petroleum system boasting high exploration success rates and low finding and development costs.

Gulfstream's assets included production from three blocks offshore Qatar and an onshore block with three natural gas fields in Oman. Of particular merit is the 19,000-acre block at the Al Rayyan field offshore Qatar, which produced 12,000 barrels of oil per day at the time of the acquisition and was ripe for significant capacity increases. The Gulfstream acquisition initially added an estimated 70 million barrels of proved oil equivalent reserves and 4,700 barrels of oil per day of production to Anadarko's worldwide portfolio.

In less than two years, Anadarko had spent more than $5 billion — most of it in stock — buying UPR, Berkley and Gulfstream. Wildly successful, the company had become a Wall Street darling, and in September 2001 its exploration prowess prompted *Fortune* magazine to pronounce Anadarko "the largest independent exploration company (and) arguably the most aggressive, most experimental and most daring oil and gas exploration company in the world." The assessment wasn't off the mark. In an industry increasingly short on cost-effective drilling opportunities, Anadarko's project portfolio stood out. The company had more drilling rigs in operation in North America than ExxonMobil at the time and was the most active driller on the continent. It listed 3,500 employees in 14 countries and possessed what *Fortune* called "one of the best balance sheets in the business, with a debt-to-total-capitalization ratio of 37 percent versus an industry average of 45 percent."

The acquisition spree had not run its course, however. In 2002, the company acquired Houston-based Howell Corporation for $220 million — funding it in part through the sale of its heavy oil properties in Canada. The four acquisitions dramatically reshaped Anadarko, giving it a global portfolio encompassing 15 million net acres domestically, 10 million net acres outside North America and five million net acres in Canada.

Jacked Up

Anadarko Qatar Energy Company is developing significant offshore assets at the Al Rayyan oil field, home to Al Morjan, the largest permanent jack-up production facility ever constructed in the Arabian Gulf. At full capacity, the production facility can process 45,000 barrels of oil a day. The experience of Sheikh Faisal Al-Thani, opposite top, a member of Qatar's royal family and veteran engineer who is "seconded," or on loan, to Anadarko from state-owned Qatar Petroleum Company, is invaluable in managing day-to-day operations. He is pictured, opposite bottom, with one of his sons and one of his prized racehorses.

Wagging Works in Wyoming

Howell Corporation owned a large stake in Wyoming's Salt Creek field, one of the largest remaining enhanced oil recovery opportunities in the lower 48 states. The century-old Salt Creek is a mature field that peaked decades ago at 90,000 barrels of oil per day and was mired at a trifling 4,000-barrel daily output, yet it was estimated to contain another 220 million barrels of oil — assuming they could be produced. Many companies had tried waterflooding the field, which failed to revive it. Anadarko's enhanced oil recovery specialists had another idea — inject the field with carbon dioxide.

CO_2 easily mixes with oil to improve its flow. Companies such as Amoco had rejected CO_2 flooding at Salt Creek because the field was low-pressure and shallow, characteristics that typically discourage its use. Based on its experience with enhanced oil recovery technology, Anadarko believed Amoco was wrong. "We first became involved with CO_2 flooding in 1994 at the Mobil GMK field in Gaines City, Texas, and picked up additional expertise participating with Exxon in Texas County, Oklahoma, and with Amerada Hess in Gaines County," explains Dane Cantwell, Anadarko's former enhanced oil recovery area manager. "We knew what we were doing and wanted a larger target to test our abilities. In Salt Creek, we found it."

As an experiment, the engineers first injected CO_2, switched to waterflooding and then returned to CO_2 — a process they call "water alternating gas" or "wagging." The tests proved highly successful. The problem was finding enough CO_2 to commercialize the field. Fortunately, they had another ingenious idea — build a 125-mile pipeline from an ExxonMobil plant in southwestern Wyoming that processed natural gas containing CO_2 from a nearby field. ExxonMobil had been stripping out the CO_2 from the gas and venting it into the atmosphere. The pipeline ended up solving the supply dilemma and benefiting the environment. Anadarko expects to sequester, or reinject, more than 23 million tons of CO_2 — a greenhouse gas — over the lifetime of the project. "We separate the CO_2 from the oil, recompress it and then use it again," Cantwell explains.

Salt Creek is the biggest CO_2 flood project in the world. Anadarko plans to produce 180 million to 220 million barrels of oil from the Salt Creek field over the next 15 to 20 years, raising production from 7,500 barrels to more than 28,000 barrels a day by 2010. "We said we could make this work, and we made it work," says Steve Pearson, currently vice president,

Oil Recovery

Paul Geiger, who came to work for Anadarko as the Salt Creek area supervisor in the Howell acquisition in 2002, contributed to the startup of the enhanced oil recovery project in the Salt Creek field near Casper, Wyoming. Anadarko is injecting carbon dioxide into this century-old field to recover as much as 220 million barrels of oil still remaining.

Good to the Last Drop

Pushing oil out of the Salt Creek field in Wyoming, which long ago seemingly had offered its last drops, put Anadarko engineers to the test. They found the solution in a process called "water alternating gas," or "wagging," in which the field is waterflooded, then injected with carbon dioxide, then waterflooded again. Production soared. Above and left, postcard views of Salt Creek in the 1950s and 1960s.

operations – eastern region. "We're always thinking out of the box here."

Salt Creek was the cornerstone of a much larger effort to expand Anadarko's oil production in Wyoming. The company applied CO_2 flooding technology to south-central Wyoming's Monell field, a Lincoln-era UPR land-grant asset in the Green River Basin. "The field produced less than 10 barrels of oil a day and was essentially abandoned," Cantwell says. "We're injecting CO_2 into the formation from the same pipeline that feeds Salt Creek and are getting about 2,500 barrels of oil out of it a day, about a third of what we project we will eventually recover."

Again, victory in one project fed success in

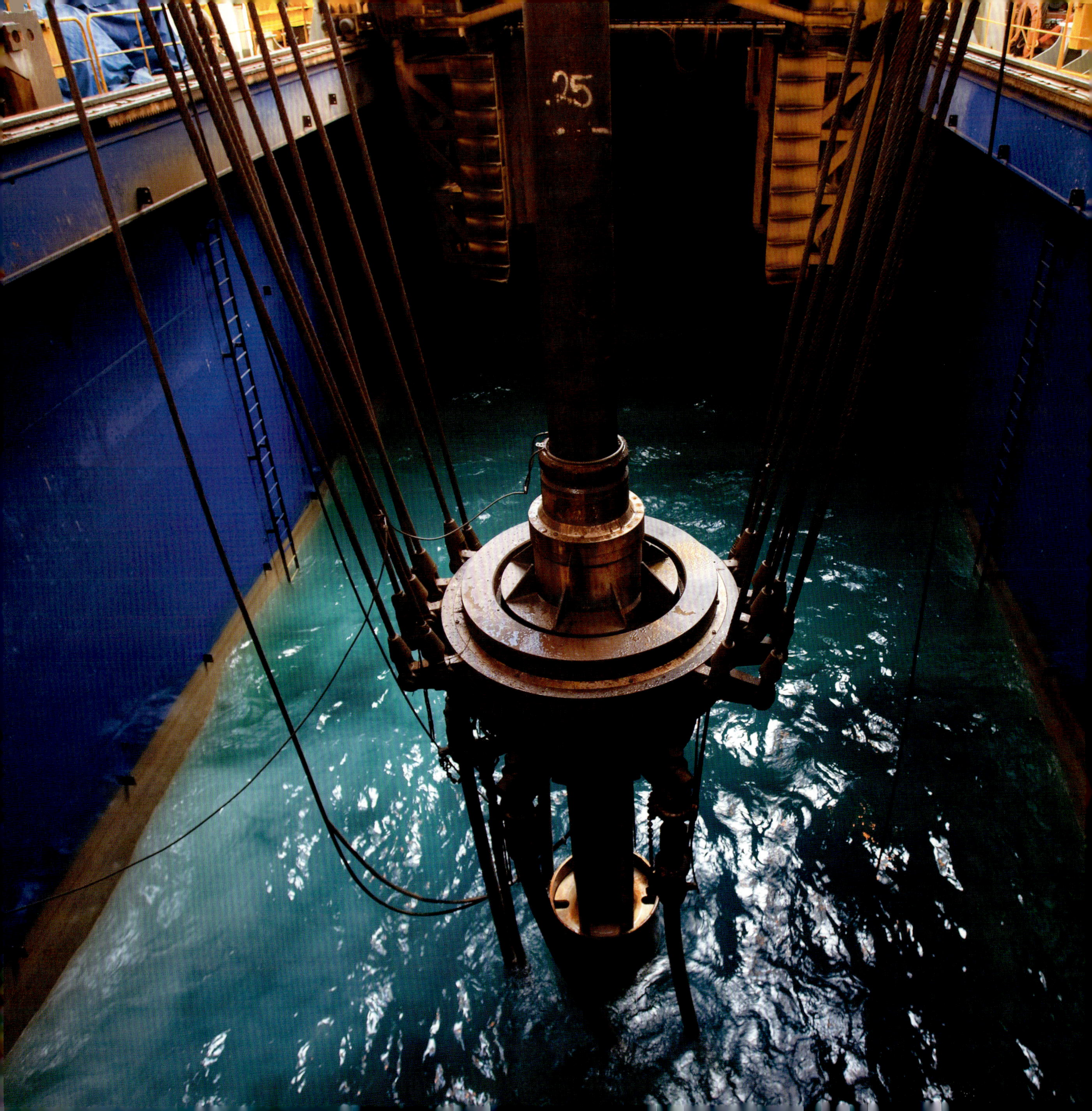
.25

Deep Water

A dynamically positioned drillship, like the Deepwater Millennium, below, keeps its drilling rig precisely on target. The moon pool, shown left, is essentially a hole in the middle of the ship to allow drilling equipment to reach the sea floor.

others. In the West Texas Permian Basin, for example, Anadarko stretched the boundaries of the Haley field. The area's over-pressured and salt-like formations had stumped geologists for decades, rendering seismic data unintelligible. Armed with new seismic interpretation technology and its singular expertise in tight gas plays, Anadarko drilled 25 development wells and two exploration wells that hiked production from zero to more than 100 million cubic feet of gas per day by the end of 2005.

Four Thousand Feet Down in the Gulf

Another frontier captured Anadarko's attention in 2000 — the deep waters of the Gulf of Mexico. Scientists had speculated for years that the Gulf's deepest reaches harbored substantial oil and gas formations. Although intrigued by the opportunity, Allison initially resisted investing in the region. "I didn't see how good sands could be deposited in deep water and I didn't think the cost curve would come down to a point where we could afford to develop it," he says. "I was wrong on both counts and I credit John Seitz for persevering in his belief that the deepwater Gulf had great potential. There are several giant hundred-million-barrel fields down there."

Anadarko made its first deepwater discovery in the spring of 2000 in the Gulf's Green Canyon area, about 160 miles south of New Orleans. Tagged Marco Polo, the find in 4,300 feet of water was a record, eclipsing the previous deepwater mark of 4,000 feet held by Shell Ursa. To develop the field, Anadarko and its partner, Enterprise Products L.P., built a mammoth floating tension-leg platform tethered to the Gulf floor by eight 28-inch-thick steel cables. It was the deepest tension-leg platform in the world at the time. The innovation did not end there. The partners

Transocean
PERSONNEL ONLY IN

In the Deep

Anadarko's deepwater exploration and production is considered a key area to its future. The company long has been active in the Gulf of Mexico, having acquired its first blocks in shallow waters offshore Louisiana in 1970. In the early 1990s, Anadarko revived exploration in the Gulf through its subsalt drilling program, proving the necessary components of an active hydrocarbon system through its advanced seismic studies. The subsalt drilling of the Mahogany well in those days was in a shallow 370 feet of water, deeper than Anadarko's early plays on the shelf but a far cry from the company's present ultra-deepwater drilling program — 9,000 feet and beyond.

The deepwater program is predicated on bringing into production reserves that lie farther and farther from existing infrastructure. Anadarko's first deepwater discovery was Marco Polo in the Gulf's Green Canyon, in 2000. At a water depth of 4,300 feet, Marco Polo set a new record. Three additional fields in Green Canyon continued the company's deepwater track record: the K2 field in Green Canyon was spud in 3,900 feet of water and drilled to a total depth of 25,700 feet, K2 North was spud in about 4,000 feet of water and reached a record 27,000 feet below the Gulf floor, and the Genghis Khan field was spud in 4,300 feet of water and drilled to a total depth of about 26,000 feet — in roughly half the time originally forecast. Anadarko's deepwater learning curve was flattening.

Now the company is exploring the ultra-deepwater regions of the Eastern and Western Gulf. In the Eastern Gulf, the company is engaged with partners in developing Independence Hub, a floating platform capable of producing an estimated one billion cubic feet of natural gas per day by mid-2007, in water depths ranging from 8,000 feet to 9,200 feet. To capitalize further on ultra-deepwater opportunities, Anadarko has locked in 15 percent of the world's deepwater rig supply with long-term contracts. "We're into it because we have the technology, the people, the capital and years of experience in the Gulf's geology, geophysics and reservoir engineering," says Stuart Strife, Anadarko's exploration manager for the Gulf of Mexico. "The next few years will be extremely exciting, if not rewarding."

State-of-the art technology defines Anadarko's deepwater exploration and production program. Like Captain Kirk aboard the Enterprise, a drilling expert at the deepwater command center, left, guides the drill bit using a joystick.

unveiled an efficient hub-and-spoke production platform — an underwater octopus into which all additional discoveries in the canyon are fed. The platform can handle 120,000 barrels of oil per day and 300 million cubic feet of gas per day, flowing from six wells as far as 15 miles away. The platform was towed to its resting place in December 2002 and came online in July 2004.

Anadarko expects Marco Polo to reach peak daily production of 80,000 barrels of oil equivalent once all the wells are online. The deepest well flows hydrocarbons beneath more than two miles of salt under extremely challenging circumstances. "Completions must be done at the bottom of the ocean, where the conditions at the sea floor are close to freezing and the circular currents wreak havoc several months out of the year," says David Jones, Anadarko senior staff geophysicist.

The project was completed on time and below budget, thanks in part to the innovative partnering arrangement. Although Anadarko operates Marco Polo, it avoided the up-front costs of building the platform, which is owned by Enterprise Products Partners and Cal-Dive International. The partners can expand their

Marco Polo

One hundred sixty miles south of the Louisiana coast lies the Marco Polo floating production platform, drawing hydrocarbons from the Earth in 4,300 feet of water. The platform has the capacity to provide 120,000 barrels of oil and 300 million cubic feet of gas per day from six wells. Jarvis Cooley, Mike McEvilly and Gary Price, left to right, opposite, were part of the team that developed the Marco Polo platform.

Filling Big Shoes
John Seitz, second from left, was the early architect of Anadarko's international exploration program — leading the company's highly successful exploration efforts in Algeria and the subsalt Gulf of Mexico. He succeeded Bob Allison as CEO in January 2002 and retired in March 2003. Seitz is pictured here with some members of his executive team: from left to right, Bill Sullivan, Dick Sharples, Rex Alman, Mike Cochran and Bruce Stover.

pipeline infrastructure to the deepwater Gulf, bolster oil and gas volumes and add to it in the future with more tie-ins to the production hub. To further hedge its risks, Anadarko financed a portion of the project, leasing the $225 million platform from Enterprise and Cal-Dive for $25 million a year plus $1 per barrel of oil recovered. "It was a great deal that made the economics work," says Mark Pease, Anadarko senior vice president of E&P technology and services.

Marco Polo motivated Anadarko to shift its focus in the Gulf of Mexico to deepwater exploration. In September 2002, Anadarko and its partners discovered the K2 field, followed 14 months later by another find dubbed K2 North. K2 was spud in 3,900 feet of water, hitting 339 feet of oil pay at a depth of 25,700 feet. At the 100 percent-owned K2 North, Anadarko drilled the deepest offshore well in its history — nearly 27,000 feet deep. Both developments have since been tied back to the Marco Polo complex. The first two wells produced a combined flow rate of 15,300 barrels of oil equivalent per day. K2 and K2 North began production in 2005 and 2006, respectively.

For more than a decade, Allison had pressed the federal government for increased access to its oil and gas resources, and he spent considerable time in Washington lobbying for the opening of a particular section in the Eastern Gulf of Mexico that had been closed to exploration since 1988. In December 2001, the Interior Department opened bids for 1.5 million acres of the area known as Lease Sale 181, and Anadarko was high bidder on 26 tracts totaling almost 150,000 acres in water depths of between 7,000 and 9,500 feet. The bids totaled $136 million.

"It's as if we've been peering over the fence for the last 13 years, with great opportunities just out of reach. Finally, the fences are coming down," a clearly pleased Allison said in a news release at the time. "We have gathered and interpreted an extensive amount of information about these blocks, and we're convinced they hold substantial oil and gas reserves with relatively low geologic risk." Indeed, in 2003 Anadarko announced its first gas discoveries in the Sale 181 area.

Allison Moves On

As all this was taking place, Allison was beginning to think about retirement. He wanted to spend more time with his family and do something other than run a major independent exploration and production company. When he informed Anadarko's board of directors, its members were taken aback. "We didn't really want Bob to retire," says John Gordon, lead director on Anadarko's board and a senior managing director at Deltec Asset Management in New York. "All of us thought it was a little too soon and none of us was happy about his decision. He was the best CEO that most of us had ever met, but we wanted to honor his personal wishes."

In January 2002, the board of directors appointed President and Chief Operating Officer John Seitz as Anadarko's new CEO. Allison remained chairman and vowed in the company's annual report to be a "very active" one at that. Seitz had distinguished himself as a proficient explorationist, earning Allison's respect for pushing forward in Algeria when others on his senior staff had wanted to pull back. Seitz also had strongly counseled the acquisition of UPR and the company's dive into deepwater Gulf of Mexico exploration. A 25-year Anadarko veteran, he had joined the company in 1977 after receiving a master's degree in geology from Rensselaer Polytechnic Institute.

Although capable and determined, Seitz had very large shoes to fill. Allison was revered by employees and heralded throughout the oil patch for his visionary leadership, an extraordinary record of success in exploration and a nimble mind with which he made decisions quickly and adeptly. Over the course of his stewardship, from 1976 when Anadarko was a subsidiary of Panhandle Eastern through its independence in 1986 and emergence as a so-called "super-independent" in the late-1990s, he had presided with integrity, passion and skill.

When Allison passed the baton to Seitz, Anadarko was at the height of its powers, posting record cash flow and earnings along with double-digit growth in reserves and production. The Bossier play and Algeria alone produced roughly 100,000 barrels of oil equivalent per day, more than the company's total daily production in 1996. Assets had grown from $750 million in 1986 to more than $20 billion in 2001, and revenues had skyrocketed in five years from $1.5 billion in 1997 to more than $8.3 billion. Best of all was the company's stock price: $75 per share, an all-time high.

Staying the Course

Bob Allison ably guided Anadarko from its days as a Panhandle Eastern subsidiary into one of the largest independent oil and gas exploration and production companies in the industry. Allison, at right, smiles alongside President Abdelaziz Bouteflika of Algeria, a country whose oil and gas reserves were crucial to Anadarko's success.

It is fitting that Allison gave the opening speech at the U.S.-Africa Business Summit in Philadelphia, where he welcomed Algerian President Abdelaziz Bouteflika, whose country had been the site of Anadarko's immense hydrocarbon discoveries in the 1990s. Bouteflika had come to America in the wake of the September 11 terrorist attacks to meet with President George W. Bush and pledge his commitment to peace. At the summit a few months later, Allison presented Bouteflika with an award honoring his work promoting the economic advancement of all Africans. Accepting the award, Bouteflika said, "Bob Allison stayed the course in Algeria when other companies left. He is a personal friend, as well as a friend of Algeria."

The Growth Curve Flattens

Anadarko's new CEO had new headquarters to go with the job. The company relocated in 2002 to The Woodlands, a more than seven-acre, bucolic campus on the banks of a man-made waterway north of Houston. The new Anadarko Tower stood 30 stories. Employees occupied the

Anadarko Tower

With Anadarko's headquarters in North Houston's Greenspoint area busting at the seams under the company's rapidly expanding workforce, Don Willis, then vice president of corporate services, suggested to Bob Allison in 1998 that the company move to The Woodlands, a thriving planned community created in the 1970s.

"Never," the chairman and CEO replied, adamant that The Woodlands — some 30 miles north of Houston — was too far from downtown.

"No obligation, but let's go see it," Willis persisted. The two went for a flyover, and Allison was intrigued by the community and its amenities. He saw that The Woodlands would offer employees a good working and living environment, which would further the company's recruiting efforts. Plus, it would provide ample room for future expansion.

But what sealed the deal for Allison was the price. "The Woodlands made us an offer we couldn't refuse," says Willis. "The brand new office space would be less expensive than what we had."

The Woodlands commercial area was in the early stages of development when, in 1998, Anadarko purchased seven and one-half acres and made plans for a 30-story tower. Anadarko's terms required the development company to locate a hotel nearby. That hotel — the Woodlands Waterway Marriott Hotel and Convention Center — can be reached from the Anadarko Tower by excursion boats that ply the man-made Lake Robbins Waterway, a floodwater retention pond that also passes a nearby shopping mall and restaurants.

Ground was broken in 1999. Less than a year passed, however, before Anadarko acquired Union Pacific Resources, followed shortly by the acquisitions of Berkley Petroleum and Howell corporations — and it was clear that Anadarko would need more space.

Coincidentally, Mobil had recently moved into the nearby Timberloch building — one day before the company's merger with Exxon was announced. Mobil abandoned its plans for its facility in The Woodlands, but the new ExxonMobil was looking for space at Greenspoint. Anadarko thus swapped its space at Greenspoint for Mobil's in Timberloch and constructed a bridge over the waterway to Timberloch.

Allison himself became deeply involved in the tower's architecture and landscaping. Testifying to Allison's love of the outdoors, the front of the tower is graced by eight bronze deer sculpted by renowned wildlife artist Dick Idol of Bigfork, Montana. Each deer is twice life-size and weighs more than 2,000 pounds. In the lobby are 10 sculptures of Canada geese with wingspans close to six feet. They were created by Ron Sweeten of Cleveland, Texas.

The tower boasts a remarkable array of amenities, including a full-service cafeteria, fitness center, a 700-person capacity meeting center named Allison Hall, a state-of-the-art graphics lab, seven IDEA (Image Delivery Enhances Awareness) rooms and a larger IDEA Theater. The IDEA facilities feature high-tech displays that provide arresting, three-dimensional views of hydrocarbon formations deep within the Earth's crust, allowing Anadarko people to extract ever more value from scientific data. These have helped geoscientists and engineers make more informed decisions about

where to drill for oil and gas deposits.

Altogether, the building sports 2,000 rooms, 8,000 windows and more than three miles of corridors.

Nearly 2,000 Anadarko people work at the tower and Timberloch. The company also fills the massive Grogan's Mill Record Center with data. This extraordinary archive contains 1.6 million seismic data logs, 200,000 nine-track tapes of additional seismic data, more than 110,000 record boxes, 75,000 geologic cores going back to the 1940s — plus the company's priceless land-grant documents, including colorful hand-drawn plats from the 1860s.

The attacks of September 11, 2001, prompted Anadarko to hang a giant patriotic banner, opposite, outside the Anadarko Tower while it was under construction. The gesture prompted hundreds of thank-yous from neighbors and passers-by on Interstate 45.

IDEA Theater

The IDEA Theater at Anadarko headquarters is where Anadarko explorationists, geologists, engineers and other scientists and technicians analyze and interpret giant color-coded seismic images, guiding them to the best places to find tomorrow's energy resources.

entire tower plus an adjacent, 11-story building called Timberloch, to which the main building is connected by a bridge.

Seitz's strategy did not veer far from his predecessor's when it came to aggressive exploration. Anadarko pursued deepwater exploration in the former Soviet Republic of Georgia, becoming the first Western company to conduct an offshore exploration program in the Georgian Black Sea. Seitz also invested in multiple projects in Australia, Gabon, Congo and Tunisia, although the bulk of the company's $500 million exploration budget in 2002 was earmarked for the Gulf of Mexico.

The collapse of one-time industry giant Enron in 2001-2002 placed a spotlight on the accounting practices of all companies. Under that glare, Anadarko missed its production forecast, distressing Wall Street. Then, in early 2002, Anadarko discovered that it had miscalculated a portion of the book value of its non-cash assets, requiring a restatement of 2001 earnings. "We had missed a calculation on a very intricate test, which required a writedown of the book value of our stated reserves," says Jim Larson, Anadarko's controller at the time. "We caught it before the end of the year but after we had issued our quarterly earnings. The timing was bad, given it was right around the time Enron was imploding," adds Larson, who later became chief financial officer before retiring in 2005.

In many respects, Anadarko was the antithesis of Enron, a company where, in the public's view, smoke and mirrors stood in for revenues and earnings. "Throughout the 1990s, as a regular part of the boardroom chitchat, there would be disparaging comments by Bob as well as others about Enron," recalls Gordon. "It was tattooed on everyone's brain that this was a company we should not emulate. When it collapsed, it was an affirmation of our thinking."

New federal accounting rules mandated in the wake of the Enron scandal by the Sarbanes-Oxley Act put pressure on corporations to upgrade financial reporting, requiring a majority of board members and all members of a board's audit committee to be independent. Both were already the case at Anadarko. Seitz moved quickly to appoint a chief governance officer, Suzanne Suter, and Anadarko's board of directors formed a corporate governance committee. The spotlight on the company soon dimmed.

A more serious threat, however, was the company's struggle to integrate its acquisitions, and this — coming at a time when Seitz had increased spending on worldwide development projects by 35 percent over the year before and finding and development costs had tripled since 1998 — had the effect of flattening Anadarko's historic growth curve. Rivals, meanwhile, were capitalizing on the high oil and gas prices to post significant gains. Anadarko's stock price plummeted in 2003 to the low-$40 range.

Fourteen months after becoming CEO, Seitz retired in March 2003. The board now had the daunting task of hiring a successor at a time when Anadarko's stock price made it vulnerable to a takeover. "Our first order of business was to find someone who could stabilize the organization and improve morale, which had become the number-one issue," Gordon says.

"The Perfect Antibiotic"

Only one person came to mind — Bob Allison. "It was an incredibly easy decision because most of us felt he shouldn't have left in the first place," Gordon says. "He was the perfect antibiotic ointment to put on the wound."

Although Allison wasn't expecting the job, he felt the call of duty. "There really was no alternative," he explains. "We needed to get this company back on track while we found a successor."

The ensuing months proved to be among the most challenging in the company's history. "We had poor operating results, poor stock performance, nagging questions about where the company was headed and speculation we'd be sold," says Larson. "The operations had become too large to manage and our costs were ballooning. Rationalization was inevitable."

The need to pare expenses required the first downsizing since Anadarko had become a public company. "It was a heart-wrenching decision, but it was something that had to be done," says a disheartened Allison, long the champion of a no-layoff policy. "We made very good settlements with people to cushion the fall, which I hope made things easier. But it was very tough to do. It couldn't have been tougher."

With morale ebbing and the corporate rumor mill buzzing that Anadarko was for sale, Allison remained focused on the hunt — not for oil this time but for the best CEO in the oil patch to lead the company through the next stage of its history. He had someone exceptional in mind and, as usual, his sights were high.

Mr. Anadarko

To many longtime Anadarko employees, Bob Allison was the human encapsulation of the company — fearless, intelligent, communicative and possessed of unimpeachable integrity. Allison, pictured here with his wife Carolyn, sometimes wore a kilt made from his Scottish family tartan to company and industry events.

CHAPTER SEVEN: SHRINKING TO GROW

Strengthening Trust

Jim Hackett was the right man at the right time to lead Anadarko. The industry veteran brought a potent combination of Harvard financial acumen, oil-patch experience and an open, frank personality well-suited to the task at hand — reorganizing and repositioning the company while strengthening trust among employees and on Wall Street.

Bob Allison sought out one of the most admired CEOs in the energy exploration and production business: James T. Hackett. Well-known in Houston's oil patch, the charismatic Hackett had high credibility with Wall Street and a reputation as a far-sighted leader. But with Anadarko a rumored takeover target and the retention of its prized geologists, geophysicists and engineers in jeopardy, would he want the job?

Hackett had spent his entire career in the energy industry and was accustomed to its inherent volatility. Long an admirer of Anadarko, and of Allison in particular, the Harvard Business School graduate was no stranger to problematic oil and gas companies. As CEO and chairman of Seagull Energy, a small independent E&P company, he guided its merger with another small independent, Ocean Energy, becoming the combined company's CEO and chairman in 1999. Integrating the companies post-merger was a formidable challenge. "We struggled with high debt levels and a desperate need to grow," recalls Bobby Reeves, executive vice president and general counsel at Ocean Energy at the time who now holds a similar position at Anadarko. "We also needed to assure Wall Street that we were balanced offshore, onshore and internationally and that employees knew what was going on and where we were going."

Hackett formed a committee of senior officers to solicit ideas on a restructuring. He then made a series of hard calls — refocusing operations, expanding exploration, spurring production and slicing costs. The resulting Ocean Energy was leaner, nimbler and focused on growth, and its stock traded at twice the pre-merger price. It was just the sort of turnaround that Anadarko's board of directors sought.

In the meantime, Ocean had been acquired by Devon Energy, and Hackett became its president and chief operating officer. In early 2003, Allison approached Hackett about taking on the CEO position at Anadarko. Hackett had reservations. "There was this sense that Anadarko was a failed company because of the layoffs and the acquisition rumors," Hackett recalls. "There also was a fair degree of skepticism in the investment community about Anadarko's reserve accounting methods and its credibility in meeting production targets. I had great respect for Anadarko and for Bob Allison, personally. The company possessed strong physical assets, tremendous intellectual capital and was the best in the independent sector in terms of technical capabilities, but I wouldn't have entertained the job were it not for Bob."

Back to Business

On December 3, 2003, Jim Hackett became Anadarko's third CEO since its independence as a public company. The 64-year-old Allison retained the title of chairman. Wall Street was bullish on the appointment: Stock analyst Irene Haas at Sanders Morris Harris issued a report commending Hackett's "solid track record" turning around energy companies and predicting that Anadarko would "get back to the business of exploration and production with a more aggressive and profit-oriented attitude."

Hackett's first official act at Anadarko was to put to rest nagging rumors that the company was on the block. "Absolutely not," he declared to the *Houston Chronicle* on his first day on the job. If the opportunity arose for a deal, Hackett

Beat the Clock

Production Engineer Jackie Hogart worked with the Wyoming County Line field team to design a new compressor unit that helped bring gas through the sales meter three months ahead of schedule. To close a four-project deepwater deal quickly, a multidisciplinary team led by Tim Trautman, exploration manager – Central Gulf of Mexico, at right above, took part in an intense three-day data-room evaluation.

insisted that Anadarko would be the acquiring company and not the other way around. Still, the handsome, 49-year-old CEO had a tough uphill climb. Although Anadarko's assets were highly regarded, overseas exploration outside of Algeria was failing to achieve acceptable returns. Furthermore, the new CEO needed to restore employees' trust and address concerns on Wall Street about the company's reserve estimates. "Jim's leadership qualities meshed perfectly with our needs," Allison says.

Over the next several months, Hackett

examined Anadarko from a strategic perspective, closely evaluating its assets, reserves, culture, organizational structure, business environment and management skills. Reserve estimation was an immediate concern, one that had prompted the board prior to Hackett's appointment to hire an outside consultant to augment the company's estimation processes. "Reserve determination is an inexact science," explains John Butler, head of the board's audit committee. "We really don't know what a field will produce until the very last drop. Nevertheless, we must do our best to ensure that the estimates are as close to the mark as possible, based on science, technology and economics."

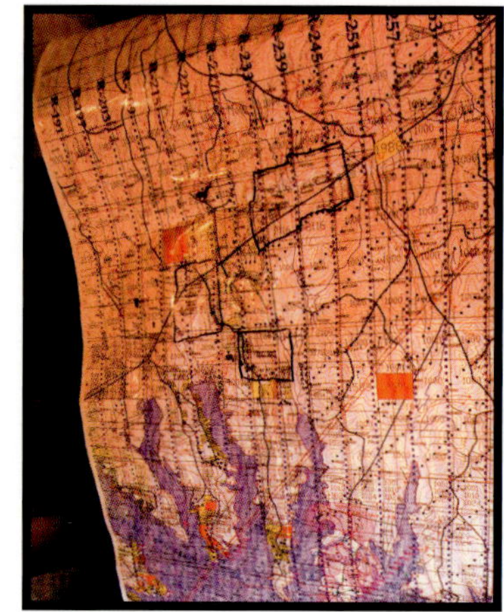

The board's actions gave Hackett comfort that "they cared about this as much as I did," he says. "I wanted a better sense of the reserve estimation process and the technical tension behind it to be sure this was something investors could depend on." The examination quickly proved the estimations reliable. "Actually, it was a more thorough process than at any other place I had worked previously," Hackett says. "The estimates were solid."

Repositioning Assets

The new CEO next attacked Anadarko's asset portfolio with the aim of strategically realigning assets to achieve sustainable growth and improved investment returns. Borrowing from his experience at Ocean Energy, he formed a committee of top-ranking executives, dubbed the Executive Committee, to assist his review. They met twice a week for nearly four months, scrutinizing global assets to identify those best suited to the growth-focused business model. As part of their evaluation, team members used a tool called Capital Planning, developed by Schlumberger to evaluate and allocate energy asset portfolios.

The team aggregated properties into 40 separate categories comprising diverse financial and operating metrics, such as operating costs, rate of return, profit margins, well-decline rates, manpower and skill requirements. The conclusion: 90 percent of Anadarko's value derived from 25 percent of its 500 fields. Properties deemed to offer less than competitive returns on capital, that had high well-decline rates and were highly capital intensive were evaluated for possible divestiture. Remaining assets were divided into two categories — "foundation" assets and "growth" assets.

The Executive Committee then stunned the industry with a bold "shrinking-to-grow" strategy in which Anadarko would dispose of a wide range of its oil and gas assets in North America. The biggest sale came in August 2004 when Anadarko sold for $1.3 billion a package of older, shallow-water Gulf of Mexico properties that the company felt it could not grow — encompassing 98.6 million barrels of oil equivalent of proved reserves and a daily net production of 46,000 barrels of oil equivalent. As Hackett explained to *Forbes*, "You can't create a competitive advantage if you're in too many places."

Anadarko also divested properties in Canada, the Rocky Mountains, the American mid-continent and West Texas. Altogether, the property

Unconventional Play

Unconventional plays remain a core competency of Anadarko during the Hackett era. Anadarko built a pipeline across central Wyoming to transport CO_2 from the Shute Creek gas-processing plant in southwest Wyoming to the company's enhanced oil recovery project in the Salt Creek field of northeast Wyoming. Opposite, a map depicts the route of a seismic acquisition project that will help Anadarko extend the boundaries of the North Louisiana Vernon tight-gas play.

dispositions represented almost 25 percent of Anadarko's production and 13 percent of reserves. In aggregate, the various divestitures generated more than $3.3 billion in before-tax proceeds. "This was unprecedented in our industry," Hackett says. "No company had ever sold off so much of itself without a merger involved. We did this not to raise cash — we had a healthy balance sheet that was strong and getting stronger. We did it because it was the right strategic decision, one that traded low-growth assets for higher-opportunity prospects." The divestitures freed Anadarko to focus its world-class exploration and unconventional resource development on its chosen assets.

Under the company's new strategic view, foundation assets were high-margin, low-risk and low-reinvestment projects with low underlying decline rates over the long term. They included 14,000 drilling locations in mature onshore basins in North America, including unconventional plays like tight gas, enhanced oil recovery, coalbed methane and fractured reservoirs in Louisiana, Texas, Wyoming and Canada, as well as the UPR land assets and remaining deepwater Gulf of Mexico properties. "We've worked hard over the past 25 years to amass a 19 million-acre position in these resource-rich areas," says Mark Pease, senior vice president of E&P technology and services.

Although the foundation assets were producing modest growth of about 4 percent a year, they generated substantial cash for investment in high-potential, high-risk growth assets and other strategic alternatives. In 2004, the foundation assets produced 116 million energy equivalent barrels, approximately 77 percent of the company's aggregate volume.

International and frontier exploration assets, most of which were slated to produce future growth, were put under the guidance of Bob Daniels, current senior vice president of worldwide exploration. The majority were in the deepwater Gulf of Mexico and the international arena. They included Algeria, Qatar, Tunisia, Indonesia and other plays deemed to possess large potential reserves and the volume growth required to evolve into stable legacy assets.

Morale Gets a Boost

To reach this ambitious target, Hackett appraised and reorganized the company's management as well as its structure. He pushed decision-making authority lower down in the ranks, improved communication and introduced employee surveys. He encouraged the implementation of a performance evaluation process called APEX, for "Achieving Performance Excellence," to inspire enhanced job performance and raise employee accountability. APEX used a scorecard system to assess individual performance and identify areas for improvement. It then linked incentive compensation to performance.

The company also invested in additional training for Anadarko's geologists, geophysicists and engineers and in state-of-the-art tools to assist them. Hackett emphasized better risk management and gave the strategic planning group a larger role in developing the asset portfolio to ensure that consistent risk-management methodologies were applied to all prospects. He also instituted a peer-review process to encourage critical discussion and collaboration.

To improve sagging morale, the new CEO espoused two-way, candid and non-job-threatening communication. He scheduled town-hall meetings to keep employees apprised of his strategy and tactics and solicited their input on areas needing improvement. When employees in Houston indicated that they would prefer to work longer hours each day if they could have every other Friday off, Hackett was willing to give the idea a try as long as productivity wasn't impaired. The program was successful and continues to this day.

Some changes breathed new life into the *esprit de corps*. Once a month in Houston (and sporadically elsewhere), Anadarko honors employees with birthdays during that month at a company luncheon. Hackett and fellow senior managers attend these events, giving presentations on the company's progress and fielding questions from the guests, who are treated to a nice lunch and birthday cake. The cultural changes had a measurable impact on morale, especially in the engineering community, where turnover dropped from 15 percent in December 2003 to 5 percent at the end of 2005.

Determined to rewrite the bottom line, the Executive Committee set out to reduce steep general and administrative expenses with an initiative to reduce controllable costs while improving various aspects of the business. Employees responded with the tactics. For example, by switching to drill bits with fewer moving parts and harder cutting surfaces, Anadarko dropped the time it takes to drill to depths of 15,000 feet to 30 days from 50 days five years earlier. This in turn lowered the cost of drilling an exploration well from $10 million to $1.8 million. In parallel with such improvements, the company pared redundant administrative activities.

"Our total capital budget was around $3 billion, but only $650 million of that flowed through the purchasing department," says Don Willis, Anadarko's former vice president of corporate services. "We'd have people in Louisiana and Texas separately negotiating with the same vendor, each getting different terms. We decided to centralize expenses through purchasing, which can negotiate the best deals." Roughly $1.5 billion, or half of G&A expenses, at the time flowed through the department.

By June 2004, after six months of strategic planning that consumed more than 10,000 hours of effort from more than 75 people across the organization, Anadarko presented its refocused

Soaring Comeback

Through the use of sophisticated technology, the Vernon field in Louisiana came alive under Anadarko. In 1999, the field produced seven million cubic feet of gas per day. By September 2005, it had set a production record of 323 million cubic feet a day. John Moran, opposite, designed a seismic data-gathering program — using "thumper" trucks, small explosive charges and compressed air guns to map what Anadarko believes will be two promising production zones at Caney Creek to extend the Vernon field.

Qatari Potential

Anadarko holds a majority operating interest in Block 12 of Qatar's Al Rayyan oil field. The Al Morjan platform, above, was built in Dubai and Sharjah in the United Arab Emirates and installed at Al Rayyan in 2003. The field also has a permanently moored tanker that functions as a storage and offloading facility. A technician, opposite, checks an electrical receptacle on the Al Morjan platform.

strategy to Wall Street. By year-end 2004, Anadarko had successfully restructured the asset base, sold more than $3.3 billion in non-strategic properties, retired $1.4 billion in debt and repurchased $1.5 billion in common stock — exceeding the Street's expectations on timing and value. Anadarko was leaner, reserve estimation was reliable, employees were more accountable, systems and processes were more modern. Morale was on the upswing. "The feedback was overwhelmingly positive, in terms of the quality and clarity of the information presented," says David Larson, who coordinated the presentations as vice president of investor relations at the time. "Almost without exception, the analysts stated that the asset realignment and new operational, financial and cultural strategies had put Anadarko on the right path."

Back in High Gear

With Anadarko on solid footing, Hackett turned his attention to the more customary tasks of running a company. Anadarko Indonesia was awarded exploration and production rights to the North East Madura III block, about 150 miles northeast of Java, in water depths of approximately 150 feet to 250 feet. The company is the operator on the block with a 100 percent working interest via a production-sharing contract with Indonesia's state-owned oil entity. "We've made a $130 million commitment to drill several wells over the next three years, and 3-D seismic surveys indicate a lot of potential," says Bob Daniels.

Another international exploration effort is underway in Qatar, which Anadarko plans to use as a base to explore additional opportunities in the Middle East. Anadarko is one of the largest Western acreage holders in Qatar, with more than 1.6 million gross acres under lease, including a 92.5 percent operating interest in the Al Rayyan oil field 90 miles northeast of Doha, headquarters of Anadarko Qatar Energy Company. The Al Rayyan field currently produces approximately 70,000 barrels of oil per day, and engineers have undertaken a comprehensive reservoir study to identify possible infill well locations to more fully develop the basin.

In Algeria, plans to construct an additional production facility are expected to boost production by another 100,000 barrels per day. Northeast from Algeria into Tunisia, Anadarko is gradually expanding its acreage position in the Berkine Basin, which it understands better than most oil and gas companies. The company also is exploring the Faroe Islands between the Norwegian Sea and the North Atlantic Ocean, where it has exploration rights in two offshore blocks spanning 617,000 acres, and has recently acquired seismic data. Another promising prospect is offshore the isle of Malta in the Mediterranean Sea. "We recently signed an agreement to join an oil and gas exploration project covering roughly 5,920 square miles in the Pilagian Basin south of Malta, which is estimated to hold 1.4 billion barrels of oil," says Richard Hook, Anadarko manager of international onshore exploration.

In the United States in 2004, Anadarko drilled 116 wells in the lower 48, finding a total of 2.2 trillion cubic feet of recoverable gas in the Vernon field alone, making it "one of the largest gas discoveries — or rather, rediscoveries — of the last decade," *Forbes* reported.

Anadarko exploration teams also hunted for hydrocarbons in frontier areas like Canada's Mackenzie Delta, a vast basin believed to contain one billion barrels of oil and nine trillion

Anadarko

Ingenuity at Work

At the Salt Creek field, a pipeline carries water produced in Anadarko's County Line coalbed methane field in northeastern Wyoming 48 miles south to Salt Creek. Injection wells then carry the water 4,500 feet below the earth's surface and push it into large underground aquifers in the Madison and Tensleep formations. Mike Blackman, right, supervised construction of the water pipeline. Brad Miller, at left, opposite, and Mike Chambers were part of the team that initiated the project. Another burst of innovation, opposite bottom, is evident at the Buckinghorse River play in British Columbia: two horizontal wells drilled on either side of the river and connected 4,600 feet below ground to enable production from the remote area.

cubic feet of natural gas. At the time, the company owned a 37.5 percent working interest in two exploration licenses covering 100,000 net acres in the area, where seismic surveyors worked in temperatures reaching 50 degrees below zero Fahrenheit.

Magnetic Attraction

Anadarko Canada also was involved in developing British Columbia's Buckinghorse River area, part of an extensive trend containing more than one trillion cubic feet of gas. The company tested a few wells north of the river to confirm substantial reserves, but the cost of running a pipeline south of the river to markets nearly doomed the project. The steep slope of the 1,000-foot gorge and stability issues at the riverbanks made a pipeline bridging the river or tunnel underneath it unfeasible.

The drilling team then had an extraordinary idea — drill two horizontal wells on either side of the river and then intersect them 4,600 feet below ground, creating a de facto pipeline north of the river to the south. To steer the wells to their intersection point, the team used a magnetic source on the drilling assembly in the south well and a magnetic receiver on the drilling assembly in the north well, and these were drawn toward each other — no easy feat since the assemblies were two miles apart. The technology provided a way to get the gas to market, and could be adapted to other resource-rich areas previously considered uneconomic.

Anadarko's redoubtable technical skills also were in evidence at the Powder River coalbed methane play in northeastern Wyoming. The water used to flood the field typically was discharged on nearby surface grounds, migrating eventually into the Powder River. Although this water is relatively pure — purer, in fact, than the river water — regulators tightened the region's water standards, requiring costly water purification. "The water we were disposing wasn't potable for humans, but was pure enough for animals and agricultural uses," says Brad Miller, Anadarko's coalbed methane general manager. "Nevertheless, it came to the point where it had become very expensive to clean up the water to the new standards."

Miller and his co-workers came up with a solution — build a 50-mile, 24-inch underground pipeline from the Powder River Basin to an underground reservoir at Anadarko's Salt Creek field, where the water would be stored for future use by area farmers at no cost to them. In early 2005, they successfully petitioned senior management for $100 million to build the pipeline. More than 400,000 barrels of water a day — enough to fill the Anadarko

Tower every five days — will be piped to the reservoir upon the project's completion in 2006. "It's a win for us and for the state, which gets as little as 10 inches of rainfall a year," Miller says.

Energized Workforce

Rising demand for oil and natural gas is energizing Anadarko's deepwater exploration and production. Since the beginning of 2004, the company has led or participated in several deepwater Gulf of Mexico discoveries including Genghis Khan, Knotty Head, Big Foot, Kaskida, Mondo Northwest, Cheyenne and Atlas NW. Most of these discoveries will be developed through the production facilities at Independence Hub or Marco Polo.

Anadarko's success in the fold belt guided ventures further east and west into even deeper waters of the Gulf of Mexico, home to its largest remaining hydrocarbon deposits. "Field sizes discovered over the last 10 years range from 50 million barrels to one billion barrels, with discoveries in the 100 million-barrel range and up," says Stuart Strife, Anadarko's vice president of exploration – Gulf of Mexico. "These ultra-deepwater wells are also more prolific, producing 5,000 barrels to 25,000 barrels a day, compared to 500 to 5,000 barrels a day from our former shelf properties."

Anadarko and partners Kerr-McGee Oil and Gas (now part of Anadarko), Devon Energy,

Field Trip

A group of Anadarko geologists and geophysicists gathers in Utah canyonland to observe, analyze and discuss rock formations and their bearing on the presence of hydrocarbons. Such field trips are part of the company's ongoing approach to education.

Mating Season

On September 20, 2006, the two pieces of the Independence Hub became one when the topsides and hull were mated to complete the semi-submersible platform in the Eastern Gulf of Mexico. The hub, operated by Anadarko, is expected to process up to one billion cubic feet of natural gas per day from an initial 15 subsea wells beginning in 2007. A lone worker descends one of the hub's staircases, opposite top. Chairman and CEO Jim Hackett, opposite below, carried on Anadarko's characteristically bold approach to growth.

Dominion Exploration and Production and Hydro formed Atwater Valley Producers Group to develop the natural gas potential in the untapped Eastern Gulf of Mexico. The companies are backing the development of Independence Hub, a floating platform capable of producing an estimated one billion cubic feet of gas per day in mid-2007 — energy vital to the U.S. supply. The design calls for a canopy of flowlines made from steel tubing to undulate from the hub several miles to 10 fields, eight of them operated by Anadarko. A 134-mile-long, 24-inch pipeline dubbed the Independence Trail will deliver gas from the hub to a pipeline owned by Tennessee Gas, which will feed it to energy-hungry customers in the United States. "The deepwater Gulf is one of our most important growth assets going forward," says Daniels.

Anadarko was at the threshold of significant growth as it approached its 20th anniversary as an independent public company. An additional 2.1 billion barrels of oil equivalent had been identified as probable in 2005, on top of proved reserves of 2.4 billion barrels of oil equivalent in 2004. Roughly half the new resource potential was related to properties under development, with the remainder attributable to exploration prospects and unconventional resource plays. Meanwhile, oil and gas prices were soaring.

Anadarko's most difficult times appeared to be behind it. Once highly valued in relation to its peers in the exploration and production sector, the company had fallen to the lower quartile in 2003. Under the new management team, Anadarko had climbed back to the middle tier in 2005 and was poised, hopefully, to regain its premier standing. Hackett's team had put its own mark on the company, making it a more modern enterprise yet one still envied for the competence of its geologists, geophysicists and engineers, the tenacity of its field crews and the dedication of its administrative staff. In November 2005, Anadarko's board put its full faith in the company's new management, naming Hackett chairman, effective January 1, 2006. In recognition of his long service and invaluable leadership, Bob Allison was named chairman emeritus.

The company entered the new year in 2006 with great confidence in its strategic plan. In short order, Anadarko would startle the oil and gas industry with a series of moves demonstrating not only a firm grasp of the company's core strengths and assets but also a bold design to bolster each.

Independence Hubbub

In the previously untapped Eastern Gulf of Mexico, or EGOM in oil and gas industry parlance, Anadarko and five partners have come together to develop a number of ultra-deepwater natural gas and condensate discoveries. Called Independence Hub, the project is based on 10 existing fields in water depths ranging from 8,000 feet to 9,000 feet — the deepest finds in the Gulf to date. The deepest well discovered so far is the Cheyenne, which set a world record in an estimated 9,000 feet of water.

A marvel of engineering, Independence Hub is a floating production platform made up of three components — the production hub, the subsea system and a 134-mile long, 24-inch export pipeline dubbed the Independence Trail. The hub itself is a deep-draft, semi-submersible platform with a two-level production deck; it's capable of processing around one billion cubic feet of gas per day. The 160 foot-tall hub consists of two parts that were mated on September 20, 2006. The hull, which gives the hub its buoyancy despite a weight of 12,000 tons, was built in Singapore. The topsides, which will provide the living quarters and production processing and compression facilities, were built in Corpus Christi, Texas. The platform and eight of the 10 discoveries so far are operated by Anadarko. The various fields will be tied back to the platform through the subsea flowline system.

While almost every aspect of the project is breaking new engineering ground, the subsea system is the star of the show. If overlaid on a map of Texas, the subsea infrastructure would reach from Anadarko headquarters at The Woodlands to Galveston, a distance of 60 miles. More than 210 miles of flowlines will bring gas from an initial 15 wells back to the hub for processing and export. Anadarko's Cheyenne discovery will be the longest tieback, at 45 miles. Following the flowlines will be 120 miles of umbilical cord, used to control wells, monitor pressures and temperatures and carry production chemicals.

Installation began in fall 2006, with the first production expected in the second half of 2007.

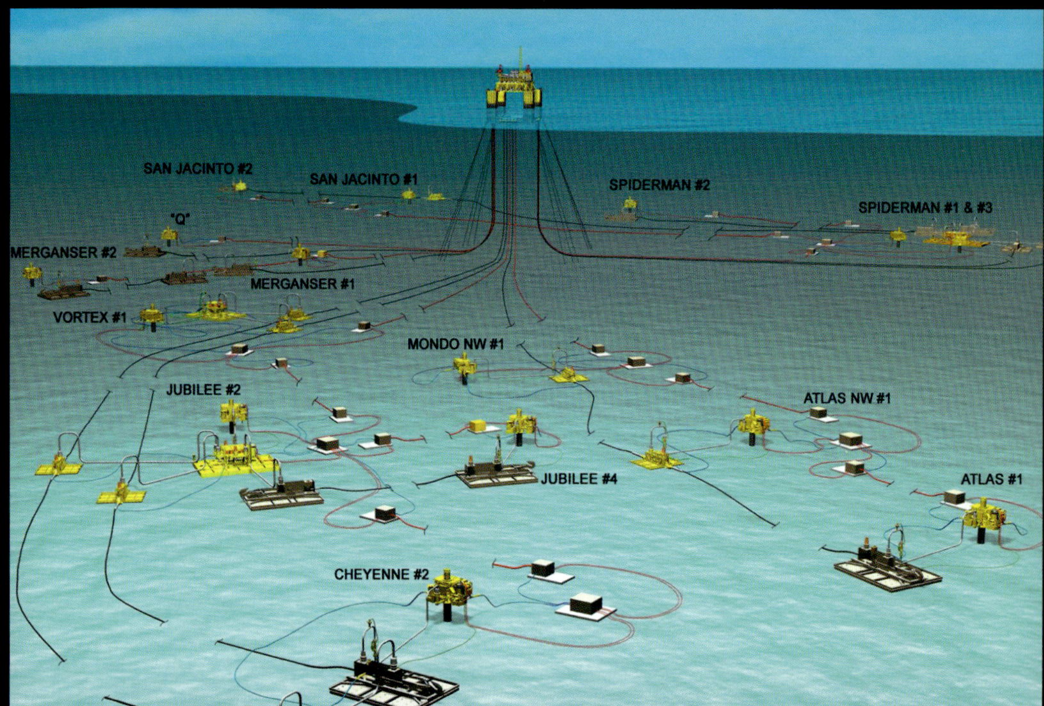

Spider Man

Anadarko has begun to develop the previously untapped Eastern Gulf of Mexico and plans to harvest gas via the spider-like Independence Hub, which was connected initially to 10 natural gas fields. Anadarko operates eight of those fields and also is the operator of the Hub itself. First production was slated for the second half of 2007. Donald LaPoint, a longtime Anadarko field foreman, was an instrumental part of the team that developed the Independence Hub platform.

Discovery in Arkansas

In 2006, Anadarko stunned the oil and gas industry by spending more than $23 billion to acquire two companies — Kerr-McGee Corporation and Western Gas Resources, Inc. Kerr-McGee enjoys a long history, dating its inception to 1929. Its first major oil discovery was the Magnolia field in Arkansas, right, in 1938.

Executive Lunch

Members of Anadarko's senior management team, known as the Executive Committee, make themselves accessible to employees, often eating lunch in the Andarko Tower cafeteria. Left to right, Karl Kurz, senior vice president, North America operations, midstream and marketing; Al Walker, senior vice president, finance and CFO; Bobby Reeves, senior vice president, corporate affairs and law; and Chuck Meloy, senior vice president, Gulf of Mexico and international operations.

The long hours spent analyzing the company's *raison d'etre* — including its assets, intellectual capital, technological capabilities and market position — yielded an incisive view of Anadarko's core competencies in unconventional resource development and high-impact exploration. Anadarko now boasted a solid North American foundation of onshore resource plays, a growing deepwater Gulf of Mexico program and an expanding international presence. Management's job was to focus the portfolio and leverage the most promising assets to fuel growth and boost future returns.

Over the years, Anadarko had often used major acquisitions and asset sales to improve its competitive position. On June 23, 2006, Anadarko engineered the boldest such move in company history — announcing a pair of acquisitions totaling $23.3 billion that would substantially increase Anadarko's size and give it industry-leading positions in two of the fastest-growing oil- and natural gas-producing regions in North America.

Following these acquisitions, Anadarko would restructure its portfolio and reduce debt through divestitures, including an agreement with Canadian Natural Resources Ltd. to sell Anadarko Canada for $4.2 billion. Excluded from the sale were Anadarko's interests in the Mackenzie Delta and other Arctic frontier properties.

The acquisition of Western Gas Resources, Inc., for $4.8 billion plus the assumption of debt took the industry by surprise. The other, an all-cash transaction to acquire Kerr-McGee Corp. for $16.4 billion plus the assumption of debt, took it by storm. Kerr-McGee was a strong competitor and a leader in resource development in the deepwater Gulf of Mexico and U.S. Rocky Mountain region — two of North America's most prolific oil- and gas-producing regions, and the geographic centers of many Anadarko plays and prospects. Kerr-McGee's proved reserves at year-end 2005 totaled approximately 900 million barrels of oil equivalent, of which around 62 percent was natural gas.

In Western Gas Resources, Anadarko acquired a valuable portfolio of natural gas gathering, processing and transportation assets in the Rockies, as well as various midstream and unconventional assets, thereby boosting its portfolio of low-risk and longer-lived tight gas and coalbed methane resource plays. Western Gas Resources' proved reserves as of year-end 2005 totaled 153 million barrels of oil equivalent.

With the closure of both acquisitions in August 2006, Anadarko became the largest U.S. producer of oil and natural gas among companies that did not also own refineries. With respect to independent exploration and production companies, it was alone in its class. Moreover, Kerr-McGee's holdings and expertise elevated Anadarko into the top echelon of deepwater operators.

Kerr-McGee was no stranger to Anadarko, having partnered with the company to back the development of the Independence Hub in the Gulf of Mexico. Renowned for its technological expertise in deepwater Gulf of Mexico exploration and production, the company had pioneered the use of offshore "spar" production platforms and had built an extensive hub-and-spoke infrastructure in the Gulf that presented significant opportunities for cost-effective future development. Core properties included more than 500 deepwater Gulf of Mexico blocks encompassing seven operated and three non-operated producing fields. The company also had three operated and five non-operated discoveries in varying stages of development in the Gulf, and four additional exploration prospects being drilled as 2006 drew to a close.

Kerr and McGee

Geologist Dean A. McGee, in hat, joined Robert S. Kerr at Anderson & Kerr in 1937 to assist the fledgling company's exploration and production efforts. McGee is credited with making Anderson & Kerr's first major oil discovery and guiding the entire industry into offshore drilling and production. The company became Kerr-McGee in 1946.

Onshore, Kerr-McGee held 451,000 net acres in the Wattenberg natural gas play in Colorado, situated largely on the Lincoln-era land grant, where Anadarko owned the royalty interest. It also held 237,000 net acres in the Uinta Basin's prolific Greater Natural Buttes gas play and other good positions and prospects, including offshore production in China, discoveries on Alaska's North Slope and offshore Brazil, and continued exploration offshore Australia, West Africa and the islands of Trinidad and Tobago. "These new assets strongly complement our existing properties, and provide the scale and focus needed to deliver more robust, predictable and efficient growth," Hackett said in announcing the acquisition.

Origins in Oklahoma

Kerr-McGee's history dates back to 1929 as the Anderson & Kerr Drilling Company, founded by Robert S. Kerr and James L. Anderson in Ada, Oklahoma. The company's modest assets in the year of the great stock market crash comprised two steam drilling rigs, three boilers, a staff of two and rising debts. By 1935, its expanding oil and gas operations required more funding than its contract drilling business could muster. To generate capital, the company was incorporated as A&K Petroleum, which then issued 120,000 shares of common stock selling at $5 per share.

With money in the bank to capitalize on booming oil and natural gas prospects in Oklahoma, A&K Petroleum searched for a skilled geologist to guide its search for fossil fuels. Dean A. McGee, then head of Phillips Petroleum's geological department, got the job and proved his merit quickly, making the company's first major oil discovery — the Barnett Number 1 well in the large Magnolia field underneath Columbia County, Arkansas. The find helped secure stronger financial footing for the company, and it boosted McGee up the corporate ladder. In 1942, he was named executive vice president of what was now called Kerlyn Oil Company.

Exploration migrated into other states,

although Oklahoma remained the primary area of focus. In 1943, Kerlyn discovered oil northwest of Oklahoma City, a major find that ignited the so-called West Edmond boom that lured oil prospectors into the region by the score. Two years later, the company broadened into the downstream segment of the oil business with the purchase of a refinery in Wynnewood, Oklahoma. It was the beginning of what would become a decades-long diversification into an integrated energy producer.

In the 1940s, Dean McGee was captivated by an extraordinary thought — drilling and producing oil in the Gulf of Mexico, beyond the sight of land. Certainly, there were no geological reasons why there could not be as much oil offshore as there was onshore, he reckoned. Yet up until this time, oil had been produced only from shallow waters. While other companies before had tried and failed to mine these resources, McGee was confident in Kerlyn's ability to make the impossible possible. "(He) just knew there was oil out there, and as good drilling contractors we knew we could do as good a job as the next man," said Kerr-McGee roughneck Otis Danielson in John Ezell's book, *Innovations in Energy: The Story of Kerr-McGee*.

Out of Sight

Kerr-McGee completed the world's first commercially productive well out of sight of land, right, in the Gulf of Mexico in 1947, effectively launching the offshore oil and gas industry. With its implementation of spar floating platforms in the Gulf of Mexico's deepest waters, the company maintained its reputation as the technological leader in offshore exploration and production.

McGee wanted to explore the Gulf's salt domes and figured the associated financial risks were more than offset by the potential of a great discovery. In 1946, after a year of promising geophysical exploration using seismic mapping techniques, the company purchased eight 5,000-acre lease blocks, located 11 miles offshore Louisiana. The following year, on November 14, 1947, on Ship Shoal Block 32, history was made — the world's first commercial oil well out of sight of land, launching the modern offshore oil and gas industry. "This is not only a huge milestone in Kerr-McGee's history, it's considered a major milestone in the industry," says Don Vardeman, former Kerr-McGee vice president of facilities engineering and now Anadarko vice president of worldwide projects.

The company had changed its name to Kerr-McGee Oil Industries, Inc., in 1946 to "identify the new man in charge," writes Ezell. In ensuing years, Kerr-McGee gradually increased its lease holdings in the Gulf of Mexico, mainly in Breton Sound, where it found oil using the first rig built for offshore work. Another industry innovation during the decade included the first semi-submersible barge for offshore drilling.

Total Energy Provider

The 1950s ushered in a new Kerr-McGee strategy to become a total energy provider. The company entered the natural gas processing business via the purchase of three plants in Oklahoma and one in Pampa, Texas. It also acquired uranium mining and milling properties in New Mexico and Arizona and more than doubled its refining capacity with the acquisition of the marketing, pipeline and refining facilities of Deep Rock Oil Corporation in Tulsa. And it bolstered its position as a distributor of refined petroleum products by acquiring Triangle Refineries Inc. in Houston. Deep Rock's gas stations in Texas and Oklahoma soon sported the Kerr-McGee brand logo.

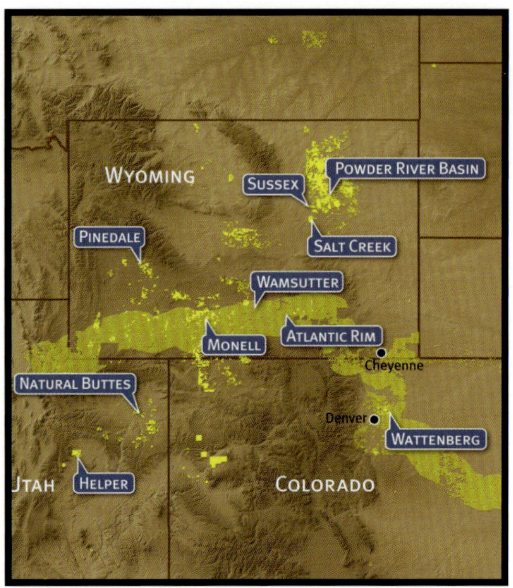

Rocky Mountains

The acquisitions of Kerr-McGee and Western Gas Resources expanded Anadarko's access to developing plays in the Rocky Mountains — among the most prolific oil and gas-producing regions in the country — along with the company's ability to gather, process and transport oil and gas to market.

McGee became the company's president and CEO in 1954. Two years later, Kerr-McGee's stock was listed on the New York Stock Exchange under the ticker symbol "KMG," and business prospects skyrocketed. In 1958, the company joined the *Fortune* 500 at number 342.

In the Gulf of Mexico, Kerr-McGee continued to lead the industry as the most technologically advanced offshore player. Its innovative rig designs included the first fully column-stabilized offshore drilling device and a rig capable of drilling in 175 feet of water while standing on the ocean floor — another breakthrough. To conduct its growing domestic and overseas offshore contract drilling business, the company formed a wholly owned subsidiary called Transworld Drilling in the 1960s.

Total energy provision remained the corporate strategy in the sixties and seventies. Kerr-McGee entered the forest products industry via the purchase of two suppliers of railroad crossties, becoming the largest producer of crossties in the country. It also entered the uranium conversion business with the startup of a facility in eastern Oklahoma.

A marked diversification away from this strategy was behind the 1967 acquisition of American Potash & Chemical Corporation, which owned a titanium dioxide pigment manufacturing facility in Hamilton, Mississippi. Titanium dioxide is a white pigment used in plastics, paint and hundreds of other consumer products. In time, the product line became one of Kerr-McGee's two core enterprises.

When weak energy prices in the 1980s nipped at the company's bottom line, Kerr-McGee abandoned its strategy of being a total energy provider. Deciding to focus on its strongest businesses, Kerr-McGee divested its interests in nuclear power, forest products and uranium. The company eventually spun off its chemical component in 2005 into a separate stand-alone company called Tronox Incorporated.

Offshore development remained a major focus of the company in the 1990s, particularly in the North Sea, where the Gryphon, Scott and East Brae fields came on stream. Gryphon is noteworthy for setting a record for fast-track field development — a blistering 10 months from government approval to first production. Other foreign offshore projects also came on-stream,

Wattenberg Play

Kerr-McGee holds 451,000 net acres in the Wattenberg natural gas play in the Rocky Mountain region, located largely on the Lincoln-era land grant, where Anadarko owns the royalty interest.

Going Deep with Spar

When Anadarko acquired Kerr-McGee in 2006, it inherited an expertise in "spar" floating production systems — boosting the company's ability to develop oil and natural gas fields in deep water. The company's spars, combined with a vast array of available deepwater technologies, are keeping Anadarko on the cutting edge of offshore exploration and development.

"In the combined company, we now have superior teams with unique skills and hands-on experience in a broad range of facility development solutions," says Don Vardeman, Anadarko vice president of worldwide projects.

Spar technology derives its name from the way it uses a mast-like, hollow metal cylinder (or, in nautical terms, "spar") as the buoyant hull of a floating platform. The cylinder can be well over 100 feet in diameter, and the platform typically is tethered to the ocean floor by cables and chain. "Imagine a very long cylinder floating vertically in water; the buoyancy on top makes it float, and the length provides stability," explains Beth Kendall, an Anadarko geophysicist who came to Anadarko from Kerr-McGee.

Prior to the use of spar technology, there wasn't an economical way to recover small to medium-sized oil deposits in deep waters with a dedicated structure. At the time of its acquisition by Anadarko, Kerr-McGee had six spar platforms in the Gulf of Mexico, as well as technological expertise in the unique production platforms.

Spar production platforms are the brainchild of Edward E. Horton, founder of Deep Oil Technology, Inc., who patented his invention in 1986. He is purported to have tested his novel concept in his backyard swimming pool.

It was 10 years before an exploration and production company took the next step — implementation — with Horton's idea. In 1996, Oryx Energy Company erected the first spar production platform, Neptune, at a water depth of 1,930 feet in the Gulf of Mexico. Two years later, Kerr-McGee acquired Oryx for $1.8 billion. "We'd been following spar technology since its inception, knew we had about 75 million barrels of oil in the field, and the development team determined our best bet was a spar production platform," says Vardeman, who managed the Neptune project.

Kerr-McGee followed up Neptune with several additional spar platforms, each more technologically advanced than the previous. The company's Nansen and Boomvang discoveries in the Gulf of Mexico utilized the world's first truss spars, which also were developed by Horton. A truss spar employs a hull that is cylindrical in shape at the upper end, with a tubular steel truss structure below. This design helps reduce overall weight and cost.

With the discovery of natural gas at its Red Hawk prospect in the Gulf of Mexico in 2001, Kerr-McGee became the first oil and gas company to utilize third-generation spar technology — the cell spar, which allows for cost-effective recovery of lower-volume commercial reserves in deep water. Red Hawk's estimated natural gas reserves of 240 billion cubic feet are 23,500 feet below the Earth's surface in 5,300 feet of water. With a cell spar, the floating structure consists of a bundle of seven rolled cylinders, each 20 feet in diameter, with one tube in the middle and six surrounding it (a cross-section of the structure looks like a grouping of cells — hence the name). "It's less labor-intensive and therefore less expensive to fabricate than more traditional spars," Vardeman explains.

The company's most recent spar platform — Constitution in the Central Gulf of Mexico's Green Canyon 680 block — is also based on the truss design. Kendall and geologist Paul Jaeger discovered the reservoir in 2001.

A Happy Upending

The first spar platform was towed to its site, opposite, and upended in 1996 in the Gulf of Mexico's Neptune field. In 2003, Kerr-McGee installed its next-generation, or truss, spar, left, at its Gunnison field. Spar platforms are well-suited for deepwater drilling because they can be moved within a field and wells can be drilled beneath them.

Kerr-McGee acquired access to it in a lease sale months later, and drilling commenced. The first well, 12,000 feet below one mile of water, was disappointing. Undeterred, the company drilled a second well, yielding a major find. "We knew there was something there," insisted Kendall, who was selected to christen the Constitution's hull at the fabrication yard in Finland.

Constitution is estimated to hold recoverable reserves of 110 million barrels of oil equivalent. First production from Ticonderoga, a second large discovery in the area that is tied back to the Constitution platform, commenced in February 2006. One month later, Constitution's resources also came online.

Spar platforms fill out Anadarko's technological portfolio for deepwater oil and gas exploration and development. The company can lay claim to six spars comprising all three generations. With these plus the Marco Polo tension-leg platform; the Independence Hub and Blind Faith deep draft semi-submersible platforms; and the floating production, storage and offloading (FPSO) unit in service in China, Anadarko is in the upper quartile of global companies in its expertise and application of facilities technology for offshore exploration and development.

including oil produced from the South China Sea and several exploratory wells drilled offshore Indonesia.

A Talented, New Company

Under new Chairman and CEO Luke Corbett, Kerr-McGee solidified its position during the decade as a major international, independent oil and gas exploration and production company. Corbett made several acquisitions, purchasing the U.K. assets of Gulf Canada Resources, the upstream U.K. business of Repsol S.A., Flag-Redfern Oil Company, Maxus Energy Canada and the domestic onshore oil and gas properties of DeltaUS Corp. In 1998, he topped them all with the $4 billion acquisition of Oryx Energy Company, which substantially increased Kerr-McGee's worldwide oil and gas exploratory prospects, reserves and production. "This was a defining moment for the company," says Darrell Hollek, former Kerr-McGee vice president of exploration and production and Anadarko's current vice president of environmental, health, safety and regulatory. "Oryx was slightly bigger than we were at the time, but was highly leveraged and in trouble. After the acquisition closed, we had a major downsizing that was difficult. But out of it came a new company boasting an extremely talented group of people."

Deepwater drilling and production was now the new frontier offshore. From its first deepwater venture in 1994 in the Pompano project in

Spar Inventor

Anadarko's Gary Mitchell, at left, and Don Vardeman, at right, talked with spar creator Ed Horton as the Gunnison truss spar was upended in the deepwater Gulf of Mexico in August 2003. At Kerr-McGee, Mitchell was Gunnison project manager and Vardeman was vice president of facilities engineering.

Deepwater Floaters

Nighttime provides a backdrop as the topside is set on the world's first cell spar platform, which came on-stream in mid-2004 in 5,300 feet of water in the Red Hawk field 210 miles south of Lafayette, Louisiana. Red Hawk is one of Kerr-McGee's six spar facilities in the Gulf; silhouetted here at sunset is another, Gunnison, whose seven subsea dry-tree wells in water 3,150 feet deep also began producing in 2004.

1,300 feet of water in the Gulf of Mexico, Kerr-McGee quickly emerged as a leader among independent oil and gas companies in this new area of development. The company earned industry renown for its use of innovative technologies, such as all three generations of spar floating production systems in the Gulf of Mexico. By reducing platform size and cost while increasing performance, spar systems offer an attractive and robust solution for developing resources in deep water. Kerr-McGee was an early adopter of spar technology and installed the world's first spar in the Gulf of Mexico in 1996, in 1,930 feet of water. The company later utilized the first truss spar, in 3,700 feet of water at the Nansen field in the Gulf, following it up with similar truss systems at the deepwater Boomvang and Gunnison fields. In 2004, Kerr-McGee achieved another industry first with its deployment of the third

generation of spar technology — the cell spar — at the Red Hawk field in the Gulf of Mexico, in 5,300 feet of water.

Amending the Constitution

The deepwater Gulf of Mexico was the site of a major Kerr-McGee find in 2001 — the Constitution field in the Central Gulf's Green Canyon 680 block. Although the first well drilled in 2001 was disappointing, geophysicist Beth Kendall and geologist Paul Jaeger, who are jointly credited with making the discovery, pressed senior management not to give up. The second well drilled in more than 5,000 feet of water turned out to be the discovery well and the field was estimated to hold 110 million barrels of oil equivalent. An additional discovery, Ticonderoga, roughly five miles south in Green Canyon block 768, forced the team to rethink the Constitution spar's capacity. It would need to handle the additional production from the nearby field. Midway through the fabrication process, the team came up with a creative solution that would increase capacity by 75 percent. As a result, the Constitution spar would be able to process 70,000 barrels of oil and 200 million cubic feet of gas a day. This additional capacity enabled the two subsea wells in the Ticonderoga field to be tied back to the spar. First production from Ticonderoga was in February 2006. The Constitution field itself came online a month later.

When invited to christen the spar in July 2005 at the Finnish fabricator before its westward cruise to the Gulf of Mexico, Kendall insisted Jaeger share the honors. "It's traditionally a woman's job to hoist the bottle, but I wanted Paul there with me," she says. "This wouldn't have happened without both of us."

It's Katrina — Turn Around

The champagne must have been flat, given what happened next. As the hull was towed from Corpus Christi, Texas, in September 2005 to its resting place 190 miles south of New Orleans, favorable weather conditions changed abruptly and drastically. "The meteorologists said they had identified something that could be a problem," recalls Mike Beattie, Anadarko general manager of international projects. It was Hurricane Katrina. "Keep in mind that the hull is 554 feet long,"

Constitutional Find

The Constitution discovery in the Gulf of Mexico's Green Canyon 680 block, above, and its sister discovery Ticonderoga five miles south marked a new beginning in cost-efficient deepwater production. Using a spar floating platform, Kerr-McGee turned an otherwise financially daunting project into a vital source of energy. Opposite, the Constitution spar awaits the installation of dry trees for its wells. These will enable the spar to gather oil from wellheads on the sea floor thousands of feet below.

Finding Tomorrow's Energy

An estimated 1.04 quadrillion cubic feet of natural gas is waiting to be discovered in the United States, and undiscovered oil off the coast of the lower 48 would provide enough gasoline to fuel an estimated 33 million cars and enough heating oil to warm 14 million homes for 60 years. Yet even as available U.S. natural gas and oil supplies dwindle and dependence on foreign suppliers escalates amid upward-spiraling energy costs, these reserves are untapped — with industry access to them denied by the government.

Obtaining access to currently off-limits energy sources is critical to the nation's energy security and economic growth, Anadarko believes. "We're facing big challenges in natural gas supply in this country," says Anadarko Chairman and CEO Jim Hackett. "There are steps that Congress and the administration can take to help us tap vast domestic natural gas resources — if we can only get permission from the government to produce them."

As 2006 drew to a close, after several years of steadily increasing energy prices, pressure to release at least some of these undeveloped resources intensified.

According to the National Petroleum Council, an advisory body to the U.S. Secretary of Energy, total U.S. demand will require the country to import 8 percent of its natural gas supply by 2008. In 2006, barely 3 percent was imported. "We're nowhere close to the ability to import what this country will need," says Greg Pensabene, Anadarko vice president of government relations.

A difference in perspective between the industry and government over access to domestic onshore and offshore natural gas reserves dates back to the environmental movement that sprang to life in the late 1960s. An oil spill off the coast of Santa Barbara, California, in 1968 catalyzed strict prohibitions on energy exploration and development offshore California. Efforts to prohibit exploration and production off the coasts of several Atlantic states were similarly successful, crushing industry hopes in the 1980s to explore significant reserves offshore North Carolina. Ultimately, the U.S. Congress put limits on exploration and production in the Outer Continental Shelf — essentially all U.S. offshore areas other than about half of the Gulf of Mexico.

With strong urging from former Anadarko Chairman and CEO Bob Allison, who stood out front for the industry in advocating the economic benefits of increased domestic oil and gas production, the government released several Eastern Gulf of Mexico tracts for exploration in 2001, via Lease Sale 181. That area had been off-limits since 1988, and Anadarko was the high bidder on 26 of the tracts. "It's as if we've been peering over the fence for the last 13 years, with great opportunities just out of reach. Finally, the fences are coming down," Allison said at the time.

They didn't come down completely, however, despite indications of intent from the Clinton and Bush administrations that the government would open the remaining Eastern Gulf of Mexico acreage to exploration. The moratoria on exploration and development on the East Coast, West Coast, half the Gulf of Mexico and offshore Alaska persisted — largely due to opposition from environmental groups concerned about possible pollution, ecological disruption and the "eyesore" of offshore development. "Production would be so far off coastlines that it could not be visible to the naked eye," says Pensabene.

Jim Hackett has taken the baton from Allison to fight for increased access, but he is not alone in this quest. Many industries outside the oil patch want the moratoria on offshore exploration lifted and the remaining Lease Sale

Gulf of Potential

Including assets picked up in the acquisition of Kerr-McGee, Anadarko holds a leading position in the Gulf of Mexico, with 2.6 million net acres and as much as 2.5 billion BOE of net resource potential. Anadarko acquired some of its most promising holdings in the government's Lease Sale 181 in 2001. But with U.S. energy demand skyrocketing, the company is pressing Congress to open up currently off-limits areas in the Eastern Gulf and other U.S. coastal regions to oil and gas exploration.

181 areas opened. In 2006, a group of CEOs that included Steve Wilson of CF Industries Holdings, a major fertilizer producer; Daniel DiMicco of Nucor Corp., the country's largest steel producer; Laurence Downes of New Jersey Resources, a natural gas distributor; Larry Nichols of Devon Energy, an independent oil and gas exploration and production company; and Hackett joined together to press Congress and the Bush administration to break the impasse "to help us increase supply, any way they can, and as soon as they can," Hackett said.

The CEOs bemoaned the impact of higher natural gas prices, a consequence of reduced supply. Downes, for example, pointed out that natural gas cost $2 to $5 per million Btu five years ago and had skyrocketed to $14 per million Btu in 2005. DiMicco said the record high prices force Nucor to pay three to five times as much for natural gas as many of its foreign competitors. Wilson said natural gas costs for the fertilizer industry were up from $80 a ton in 2000 to nearly $500 a ton in 2005, putting a strain on American farmers. "We either abdicate to emotional, uninformed legislation and regulation or fight the good fight to advance the world's health and welfare in an intelligent and evolutionary manner," Hackett asserts.

In September 2006, as Congress prepared to adjourn for the election season, House and Senate negotiators raced the clock to reach an agreement on opening the Outer Continental Shelf to exploration and development — but no deal was reached, and the issue remained for another Congress to address.

Hot Prospect

Since around 2000, Anadarko has considered coalbed methane, which involves extraction of natural gas from thick coal deposits, a core strategic play. Western Gas Resources, which Anadarko acquired in 2006, had already been acquiring and developing coalbed methane holdings on its way to becoming the second largest leaseholder in one of the nation's most prolific fields — the Powder River Basin of northeastern Wyoming and southeastern Montana, above. Western Gas assets were directly adjacent to those of Anadarko, and as a result of the acquisition Anadarko is now one of the basin's largest asset holders, with 722,000 net acres. A solitary worker, right, coaxes gas from the basin's coalbed methane reservoirs.

Beattie says. "Turning it around quickly to get back to port was logistically challenging, requiring regulatory coordination and quick approvals. It was a mighty effort and, thankfully, we made it back intact."

After the storm subsided, the spar was successfully installed, its mooring wires tethered firmly to the ocean floor. "No sooner was the platform declared storm-safe than Hurricane Rita came close on our heels, just missing us," Beattie says.

At the start of production, Constitution yielded about 15,000 barrels of oil and 12 million cubic feet of natural gas per day, while the nearby Ticonderoga field was producing approximately 22,000 barrels of oil and 18 million cubic feet of natural gas per day through the spar.

Western Gas: Strength Midstream

While the Kerr-McGee acquisition was the bigger of the two announced in 2006, the purchase of Western Gas Resources is equally important to Anadarko's long-range future. The company is a major player in two natural gas resource plays in Wyoming — coalbed methane in the Powder River Basin and tight gas in the Pinedale Anticline, with the coalbed methane properties alone estimated to hold about nine trillion cubic feet of gas.

Western Gas Resources dates its beginnings to 1971 as Ecological Engineering Systems, Inc., founded by Brion Wise, Walt Stonehocker and two others. Recognizing the need to capture natural gas flaring from producing wells, the Denver-based company forged a niche providing natural gas gathering, processing and marketing services to producers. It quickly developed a strong reputation for designing and constructing complex gas-processing facilities and providing low-cost natural gas services. In 1977, it changed its name to Western Gas Processors.

In 1979, the company constructed the large Teddy Roosevelt sour gas treating facility in North Dakota. "This was the event that put the company on the map," says Dave Keanini, former vice president of engineering at Western Gas Resources and now Anadarko general manager of midstream engineering for the Rockies. "It was much bigger than anything else we had built, and we began to grow steadily." Indeed, only 11 employees were on the payroll just after the plant was erected. Seven years later, the number of workers had increased to 200.

The 1980s were a period of sustained growth for Western Gas Processors. It built numerous plants, including Baker in Montana, Amos Draw and Lincoln Road in Wyoming, Four Corners in

Patrick Draw and Durham Ranch

Western Gas Resources has earned a reputation for deploying de-watering technology to release trapped natural gas in the Powder River Basin's coalbed methane play. At left is the company's Patrick Draw gas-processing plant. As with other assets, maintaining environmental integrity was a high priority in Western Gas' development of another coalbed methane play, the Durham Ranch, above.

Utah, and the Williston and Temple plants and Alexander gathering system in North Dakota. In 1985, when Western Gas acquired MCO Companies, five more processing plants in the Powder River and Williston basins were added to the company's portfolio.

Going Public for Growth

To fund a five-year, $100 million capital expenditure program, Western Gas Processors became a public company, and its stock began trading on the New York Stock Exchange in 1987. "Our goal was to expand our regional presence, as well as grow our gathering and processing business," says Teresa Perry, former director of corporate communications and services at Western Gas and now an Anadarko corporate communications advisor. "Our expertise in providing midstream services to third-party producers was providing many opportunities for us to expand our operations."

The capital influx funded the acquisition of several plants around the country, beginning in 1989 with the Midkiff plant in West Texas and the San Juan River plant in New Mexico. The campaign culminated in the purchase of Union Texas Petroleum's gas processing assets in 1991, the year the company changed its name to Western Gas Resources. The acquisition added the Chaney Dell, Benedum and Toca plants in the states of Oklahoma, Texas and Louisiana to the growing asset portfolio and doubled overnight the number of wells, dedicated reserves, gas throughput, gathering miles and workforce. Western continued to grow its asset base in the 1990s with the purchase of Mountain Gas Resources and the Granger and Red Desert facilities in Wyoming. It also built the Katy Gas Storage and Bethel Gas Treating facilities in Texas.

Midkiff Sunset

The sun sets on Western Gas Resources' Midkiff gas plant in West Texas.

In 1996, revenues catapulted to $2 billion and the number of employees neared 1,000. Under new President Lanny Outlaw, Western Gas significantly expanded its presence in the Powder River Basin, one of the most bountiful natural gas fields in the United States, with an estimated 25 trillion cubic feet of gas. The company signed an agreement with Barrett Resources (Williams Companies today) to develop 2.1 million acres in the basin, becoming the largest acreage holder and producer in the basin. It also inked an agreement with Ultra Petroleum to develop an additional 1.8 million acres in southwest Wyoming's over-pressured, tight gas Pinedale field, another promising natural gas resource.

Western Gas Resources' Wyoming assets were a major factor in Anadarko's determination to buy the company. "Together, we have more than 12,000 identified drilling locations in inventory," Hackett explains. "By adding our adjacent properties to theirs, we can accelerate the development of these combined resources, producing strong growth through the decade and possibly longer."

"People Here Do Not Give Up"

The twin acquisitions of 2006 brought to the fore a new Anadarko, now an upper-tier player in exploration and production in the deepwater Gulf of Mexico and Rocky Mountains and one with new opportunities to develop resources to provide energy for America.

Anadarko's 5,300 employees around the world had much to cheer in 2006. Not only had their company survived the exigencies of a challenging period intact, prosperous and steadily on course, it had shored up its core strengths to confront the demands of the future. Still, it is the human side of their work at Anadarko that many find most compelling. Asked to name what they like best about their work, few Anadarko employees mention the big plays like Algeria and the deepwater Gulf of Mexico or the satisfaction of pumping hydrocarbons from long-dormant fields. "Friends in the oil patch ask why I've stayed so long at Anadarko," says Steve Martin, whose innovative waterflooding ideas show the inventive spirit that still permeates the company. "I tell them I'm here not because of the assignments, which are challenging and rewarding, but because of the people I work with, day in and day out," he adds. "I respect their skill, their intelligence and their humor. It is a joy to come here."

Others express similar feelings of respect, friendship and camaraderie. "The people around me are extremely intelligent and creative, truly in the top percentile of intellects — present company excluded," laughs Julie Struble, one of the first female engineers employed at Anadarko. "In the 24 years I've been here, I've received headhunter calls I don't know how many times. I have never met with anyone outside this company to discuss a job because I knew that, whatever they told me, it would not compare to Anadarko. The people I work with, and who support me to do more and go further, knock me off my seat. I'm in the middle of a great ride."

Midstream Key

As manager, regulated pipelines, Christine Odell plays an important role in overseeing sales and transportation through regulated pipelines and ensuring compliance with state and federal regulations.

Many employees praise former CEO Bob Allison's stalwart leadership through good times and tough ones, along with Hackett's determination to reshape the company.

"People here do not give up," says Rex Alman, who doggedly promoted the production-generating Sudden Impact project that helped Anadarko meet its 1997 production targets. "Employees kept coming up with ideas, plans and projects that never officially made the annual report, but they're there in droves," he explains. "Bob Allison made us focus on external competition through exploration, and he established a corporate culture of always doing the right thing. His message was always, 'Rex, if you have to make a quick decision and can't reach me because I'm on an airplane or something, just do the right thing. Do the right thing.' I've lived my life by that advice."

Hackett similarly espouses ethical decision-making. "Jim is all about integrity," says Bobby Reeves, the lawyer who followed his boss from Ocean Energy to Anadarko. "His focus is on sustainable commercial success and building trust with all stakeholders. Anadarko is still based on the same family values that have characterized its history. People are the cornerstone of this company."

When Anadarko executives today recruit the people of tomorrow, they target only those who share their corporate values. While Anadarko may not currently have the same household-name value as its big-oil rivals, it rarely has trouble finding great talent. "We have no qualms about competing against larger, integrated oil and gas companies for the best candidates," says geophysicist Robert Talley, whose seismic studies of Matagorda Island led to Anadarko's great offshore discovery in 1980. "Why should we?"

Years ago, Talley says, students would ask, "Who is Anadarko?" "Those days are long over," he smiles, his eyes narrowing to make a point. "Today, they not only know us, they want to be us."

TIMELINE

1928
Missouri-Kansas Pipeline Corporation is founded as a holding company with one income-producing asset, Panhandle Eastern Pipe Line Company, which develops and builds a major pipeline from Kansas to Indiana.

1943
Panhandle Eastern is established as an independent company.

1959
Panhandle Eastern creates Anadarko Production Company as an exploration and production subsidiary. Anadarko, headquartered in Liberal, Kansas, spuds its first well.

1965
Anadarko acquires the assets of Ambassador Oil Company for $12 million. Ambassador's holdings span 600,000 acres in 19 states and Canada.

Anadarko headquarters relocates to Fort Worth, Texas, the former headquarters of Ambassador.

1970
Anadarko takes first steps towards offshore exploration by joining forces with five other companies to acquire drilling rights to nine blocks in the Gulf of Mexico.

1974
First Anadarko production from the Gulf of Mexico begins at Vermilion block 320.

Anadarko headquarters moves from Fort Worth to Houston.

1975
Anadarko substantially increases its ownership of acreage in the Gulf of Mexico, acquiring an interest in 150 blocks spanning 74,000 acres.

1976
Anadarko gains control of Panhandle Eastern's Pan Eastern Exploration and Pan Western Exploration subsidiaries as part of Panhandle's decision to achieve competitive natural gas pricing on its older properties.

1978
Anadarko surpasses $100 million in gross revenues.

1979
Anadarko discovers hydrocarbons at the High Island A-376 field off the Texas coast.

1980
Anadarko headquarters moves to Greenspoint in North Houston.

Company records its first major offshore success in the Gulf of Mexico at Matagorda Island block 622/623. Production begins in 1984, and by the end of 1998 more than 899 billion cubic feet of gas and 9.3 million barrels of oil flow from the field.

1986
On September 8, Anadarko begins trading on the New York Stock Exchange. On October 1, Anadarko is spun off to Panhandle Eastern stockholders, officially marking the birth of Anadarko Petroleum Corporation as an independent company. Proved reserves of 313 energy equivalent barrels ranks Anadarko as one of the nation's largest independent exploration and production companies.

Kansas Corporation Commission approves infill drilling in the Hugoton Embayment, a decision that will more than double Anadarko's gas production in the area.

1987
Anadarko is ranked 19th in the United States and 20th in the world for natural gas reserves.

1988
The company extends the productive limits of the Matagorda Island block 622/623 field in the Gulf of Mexico with success at the Matagorda Island block 635 No. 1 well.

1989
Anadarko lands its first major international exploration deal by signing a production-sharing agreement with Sonatrach in Algeria.

In response to Federal Energy Regulatory Commission Order 451, Anadarko cancels low-price gas contracts with its largest customer, Panhandle Eastern, and begins selling gas at higher prices to clients nationwide.

Anadarko acquires 29,000 acres in the Arkoma basin.

1990
Anadarko establishes its reputation as a technology-driven company by becoming one of the first among oil and gas exploration and production companies to use 3-D seismic workstations in-house. The company also develops prowess in using unconventional processes like hydraulic fracturing at the Golden Trend play in Oklahoma.

Hugoton Gathering System (HUGS) is completed, giving Anadarko greater control of its natural gas assets in Kansas.

1991
Anadarko's first Algerian well is spudded — a dry hole. Nevertheless, the Berkine West No. 1 well provides valuable information that points geologists toward more promising prospects in the region.

Anadarko successfully explores formations deep beneath the Hugoton field's traditional shallow pay zones.

1992

Anadarko invests more than $200 million to purchase properties in the Permian Basin, tripling its production in the area.

Anadarko headquarters moves to Anadarko Tower in Houston's Greenspoint area.

1993

Anadarko and its partners usher in the era of subsalt exploration in the Gulf of Mexico with the discovery of the Mahogany field.

Anadarko announces the discovery of the El Merk No. 1 well in Algeria.

1994

Anadarko and ARCO Alaska discover the Alpine field on the North Slope of Alaska.

1995

Anadarko discovers two giant fields in Algeria — Hassi Berkine South and Ourhoud.

1996

Anadarko drills first successful wells in the Bossier field in Freestone County, Texas.

1997

"Sudden Impact" project increases year-over-year domestic production by 18 percent.

1998

Production from the Hassi Berkine South field in Algeria commences.

Anadarko strengthens its position in Alaska by signing an exclusive agreement with the state's largest native corporation to explore more than 2.2 million acres in the foothills of the North Slope.

2000

Anadarko merges with Union Pacific Resources, solidifying its position as one of the world's largest and most successful independent exploration and production companies. The combination doubles Anadarko's proved reserves and gives the company expanded holdings in Canada.

Company announces its first discovery in the deepwater Marco Polo field in the Gulf of Mexico. Deepwater will become the growth engine of Anadarko's offshore program.

2001

Anadarko acquires Canada-based Berkley Petroleum Corporation.

Company buys Gulfstream, a Canadian company with assets in Qatar and Oman.

Anadarko joins the *Fortune* 500. It is listed as the fifth largest holder of natural gas reserves and the fifth largest gas producer in North America.

2002

Anadarko acquires Howell Corporation, giving it access to Wyoming's Salt Creek field, one of the largest remaining enhanced oil recovery opportunities in the lower 48 states.

Anadarko announces the K2 discovery in the Gulf of Mexico.

Anadarko moves into new headquarters at Anadarko Tower in The Woodlands, Texas.

2003

Anadarko announces the first of several natural gas discoveries in the Lease Sale 181 area of the Eastern Gulf of Mexico.

Anadarko announces a discovery on block 404 and gains approval to develop block 208 in Algeria.

The Al Morjan platform is commissioned in Qatar.

2004

Anadarko unveils a new corporate strategy to focus on higher returns and sustainable annual production growth. The strategy involves shedding some $3 billion in properties and an intense focus on assets to determine which are the company's foundation and which are capable of long-term growth.

Company is awarded exploration and production rights to North East Madura III block in Indonesia.

The Al Rayyan oil field in Qatar becomes the base of Anadarko's Middle East growth platform.

The Wild River play becomes a core producing area for Anadarko in Canada.

2005

Company announces Bear Head project in Nova Scotia to receive, store and regasify liquefied natural gas from the Atlantic Basin.

Anadarko announces the Genghis Khan deepwater discovery in the Gulf of Mexico and begins exploring the Green Canyon in the eastern Gulf of Mexico.

2006

Installation of the Independence Hub in the Gulf of Mexico's deep waters begins.

Anadarko announces the Kaskida oil discovery in the Gulf of Mexico — the industry's first discovery in the lower Tertiary. Kaskida encounters 800 feet of net pay and sits in about 5,800 feet of water.

The company is awarded rights to explore and produce offshore Mozambique in a 2.64 million-acre tract.

Anadarko acquires Kerr-McGee Corporation and Western Gas Resources, giving Anadarko industry-leading positions in two of North America's fastest-growing oil and gas producing regions — the deepwater Gulf of Mexico and the Rockies.

Anadarko agrees to sell its Anadarko Canada subsidiary, minus interests in the Mackenzie Delta and certain Arctic frontier properties, to Canada Natural Resources Ltd. for $4.2 billion.

ACKNOWLEDGMENTS

I would like to thank the many Anadarko employees and retirees who gave their time so fully to me in writing the history of this great company. Dozens of interviews were conducted, and several retirees provided crucial information and memorabilia that I could not otherwise have obtained.

Two books were important in providing a historical perspective of Anadarko's founding — *Summer of '86* by Tom Irwin and *Gas Pipelines and the Emergence of America's Regulatory State* by Christopher Castaneda and Clarance Smith. I also want to acknowledge historian Maury Klein's *Union Pacific*, a book that provided invaluable insight into the story of Union Pacific Resources.

Many thanks also are due Teresa Wong, Paula Beasley and Anadarko's media relations staff in Houston and Calgary for their interview, research and archival assistance — not to mention unflagging zeal in this project. I'd also like to acknowledge the fine work of my assistant Jennifer Sue Johnson, whose many hours organizing the research data were instrumental to the book's completion.

ABOUT THE AUTHOR

Russ Banham is a veteran business journalist and the author of 10 books, including the best-selling *Rocky Mountain Legend*, the story of the Coors brewing dynasty; *The Ford Century*, the award-winning centennial history of Ford Motor Company; and *Wanderlust: Airstream at 75*. His television appearances include NBC's *Today* show and *Biography* on the A&E channel.

INDEX

Bold listings indicate illustrations.

A

B

C

D

E

F

G

H

I

J

K

L

M

N

O

P

Q